T0302243

Materials and Devices for End-of-Roadmap and Beyond CMOS Scaling

MATERIALS RESEARCH SOCIETY
SYMPOSIUM PROCEEDINGS VOLUME 1252

Materials and Devices for End-of-Roadmap and Beyond CMOS Scaling

Symposium held April 5–9, 2010, San Francisco, California

EDITORS:

Shriram Ramanathan
Harvard University
Cambridge, Massachusetts, U.S.A.

Andrew C. Kummel
University of California—San Diego
San Diego, California, U.S.A.

Supratik Guha
IBM Thomas J. Watson Research Center
Yorktown Heights, New York, U.S.A.

Heiji Watanabe
Osaka University
Osaka, Japan

Jochen Mannhart
Center for Electronic Correlations and
Magnetism
Institute of Physics
University of Augsburg
Augsburg, Germany

Iain Thayne
University of Glasgow
Glasgow, Scotland, U.K.

Prashant Majhi
Sematech/Intel
Austin, Texas, U.S.A.

Materials Research Society
Warrendale, Pennsylvania

CAMBRIDGE
UNIVERSITY PRESS

University Printing House, Cambridge CB2 8BS, United Kingdom

One Liberty Plaza, 20th Floor, New York, NY 10006, USA

477 Williamstown Road, Port Melbourne, VIC 3207, Australia

314-321, 3rd Floor, Plot 3, Splendor Forum, Jasola District Centre, New Delhi - 110025, India

79 Anson Road, #06-04/06, Singapore 079906

Cambridge University Press is part of the University of Cambridge.

It furthers the University's mission by disseminating knowledge in the pursuit of education, learning and research at the highest international levels of excellence.

www.cambridge.org
Information on this title: www.cambridge.org/9781605112299

Materials Research Society
506 Keystone Drive, Warrendale, PA 15086
http://www.mrs.org

© Cambridge University Press 2010

First published 2010
First paperback edition 2012

Single article reprints from this publication are available through University Microfilms Inc., 300 North Zeeb Road, Ann Arbor, MI 48106

CODEN: MRSPDH

A catalogue record for this publication is available from the British Library

ISBN 978-1-605-11229-9 Hardback
ISBN 978-1-107-40798-5 Paperback

CONTENTS

*Invited Paper

POSTER SESSION

III-V MOSFET

NOVEL DEVICES AND III-V MOSFET

MATERIALS AND DEVICES FOR BEYOND CMOS SCALING

PREFACE

This proceedings volume contains papers presented at Symposium I, "Materials for End-of-Roadmap Scaling of CMOS Devices," and Symposium J, "Materials and Devices for Beyond CMOS Scaling," held April 5–9 at the 2010 MRS Spring Meeting in San Francisco, California. These symposia attracted 106 presentations, of which 22 were invited.

Historically, scaling in Si CMOS was primarily led by lithography. In the last decade, this situation has been completely revolutionized with the introduction of the likes of copper interconnects, high-k gate dielectrics, metal gates, and strained silicon to meet the demands of the International Technology Roadmap for Semiconductors as the technology generations were reduced beyond 45 nm. As we look towards the end of the roadmap and beyond, the proliferation of potential solutions to meet the necessary performance challenges becomes truly staggering, and has motivated an exponential increase in research in a wide range of emerging materials and devices architectures.

This volume contains the refereed versions of numerous papers presented in the symposia which collectively capture the diversity of research activity being actively pursued around the world to address the very significant challenges faced at the end of the CMOS Roadmap and beyond. We believe the papers in this volume are a very revealing "snapshot in time" of the state of research in this dynamic field, and that this collection will be of significant benefit to researchers in the area, both now and in future, as a valuable reference.

The organizers are indebted to the speakers for their interest and support of the symposia by submitting such high quality abstracts and excellent, captivating presentations and posters.

We are very grateful to the staff of the Materials Research Society, who was of enormous assistance at all stages before, during, and after the symposia and during the publication period.

Andrew Kummel
Heiji Watanabe
Iain Thayne
Prashant Majhi
Supratik Guha
Jochen Mannhart
Shiram Ramanathan

November 2010

MATERIALS RESEARCH SOCIETY SYMPOSIUM PROCEEDINGS

MATERIALS RESEARCH SOCIETY SYMPOSIUM PROCEEDINGS

Volume 1270 — Organic Photovoltaics and Related Electronics—From Excitons to Devices, V.R. Bommisetty, N.S. Sariciftci, K. Narayan, G. Rumbles, P. Peumans, J. van de Lagemaat, G. Dennler, S.E. Shaheen, 2010, ISBN 978-1-60511-247-3

Volume 1271E —Stretchable Electronics and Conformal Biointerfaces, S.P. Lacour, S. Bauer, J. Rogers, B. Morrison, 2010, ISBN 978-1-60511-248-0

Volume 1272 — Integrated Miniaturized Materials—From Self-Assembly to Device Integration, C.J. Martinez, J. Cabral, A. Fernandez-Nieves, S. Grego, A. Goyal, Q. Lin, J.J. Urban, J.J. Watkins, A. Saiani, R. Callens, J.H. Collier, A. Donald, W. Murphy, D.H. Gracias, B.A. Grzybowski, P.W.K. Rothemund, O.G. Schmidt, R.R. Naik, P.B. Messersmith, M.M. Stevens, R.V. Ulijn, 2010, ISBN 978-1-60511-249-7

Volume 1273E —Evaporative Self Assembly of Polymers, Nanoparticles and DNA , B.A. Korgel, 2010, ISBN 978-1-60511-250-3

Volume 1274 — Biological Materials and Structures in Physiologically Extreme Conditions and Disease, M.J. Buehler, D. Kaplan, C.T. Lim, J. Spatz, 2010, ISBN 978-1-60511-251-0

Prior Materials Research Society Symposium Proceedings available by contacting Materials Research Society

Novel Devices

Mater. Res. Soc. Symp. Proc. Vol. 1252 © 2010 Materials Research Society   1252-I02-04

Tunneling MOSFETs Based on III-V Staggered Heterojunctions

P. M. Asbeck[1], L. Wang[1], S. Gu[1], Y. Taur[1] and E. T. Yu[1,2]

[1]Electrical and Computer Engineering Department, University of California, San Diego, 9500 Gilman Drive, La Jolla, CA 92093, USA

[2]Present address: Microelectronics Research Center, University of Texas, Austin, TX 78758, USA

ABSTRACT

A critical problem for the progression of CMOS electronics to the nanoscale is the reduction of power density, while at the same time preserving high speed performance. One of the most promising approaches is to aggressively reduce the power supply voltage by using a novel device, the tunneling MOSFET (TMOSFET), which is a MOSFET that operates by tunnel-injection of carriers from source to channel, rather than by conventional thermionic emission. TMOSFETs benefit from steep (sub-60mV/dec) gate turn-on characteristics. In this paper we show that TMOSFET designs based on staggered heterojunctions are particularly promising, since the choice of materials for the injector (source) and channel allows optimization of the tunneling probability at the heterojunction. Analysis and simulation of MOSFETs based on the GaAlSb / InGaAs material system are presented. The energy offset between the valence band of the injector and the conduction band of the channel at the heterojunction can be tailored over a wide range, from negative values ("offset" band lineup) to values in excess of 1eV. We find by simulation that for optimal values of effective heterojunction bandgap near 0.2eV, the resulting MOSFETs are capable of delivering >0.5mA/mm while maintaining on-off ratio greater than 10^4 over voltage swing of 0.3V. We also discuss a variety of materials-related challenges that must be overcome to realize the predicted performance. Among these are the need to provide near ideal heterojunctions between the materials, employ high K dielectrics with very low interface state density, and achieve good alignment between the gate and the heterojunction. Different configurations for the tunneling MOSFETs are presented.

INTRODUCTION

During decades of progress in CMOS electronics following Moore's law, the power dissipation per unit area has been kept within bounds by a progressive scaling of the power supply voltage V_{DD}. Further reduction in V_{DD}, however, is impeded by the finite sub-threshold slope associated with MOSFETs of conventional design. It is well-known that MOSFET drain current varies with gate voltage at best at a rate of 60 mV/decade at room temperature (and often more slowly) because the transistor operation relies on thermal excitation of carriers over an energy barrier (modulated by the gate voltage) between the source and the channel [1]. In order to maintain a low OFF current in a typical MOSFET, the threshold voltage cannot be lower than about 0.25-0.3 V. To achieve a useful ON current, a gate overdrive voltage of order 0.3-0.5 V is also required, dictating that values of V_{DD} for high performance applications cannot be decreased below 0.6-0.8V. Since power dissipation due to switching energy is proportional to $C_{eff} V_{DD}{}^2 f_{clk}$, where C_{eff} is an effective load capacitance and f_{clk} is the clock frequency, it is difficult to avoid increasing power density per unit area as gate dielectrics shrink and f_{clk} increases.

An approach to addressing the power dissipation problem has recently received considerable attention, based on Tunneling MOSFETs [2-9] (abbreviated here as TMOSFETs). These devices are expected to attain sub-threshold swings considerably better than conventional MOSFETs, down to the 20 mV/decade region, as shown in Fig. 1. As a result, it can be anticipated that power supply voltages for CMOS electronics could be reduced to values as low as 0.2-0.25V while still maintaining high speed performance. In order to achieve this goal, the TMOSFETs should, in addition to the sub-threshold swing metric, achieve low values of OFF current (of order 10-100 nA/um of gate width) and high values of ON current (of order 1mA/um of gate width).

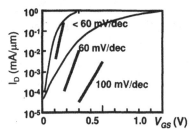

Figure 1: Schematic I_d-V_{GS} characteristics of present-day MOSFETs, and those of proposed TMOSFETs.

The fundamental operating principle of the TMOSFET is to provide current from the source to the channel by a tunneling process rather than by thermionic injection. This permits the output current to be independent of KT, and thus avoids the 60 mV/decade constraint. Figure 2 compares the schematic band diagram of a conventional MOSFET and that of the TMOSFET under operating conditions. Carriers tunnel between the valence band of the source and the conduction band of the channel (for an n-channel device).

One of the critical considerations for the implementation of high performance TMOSFETs is to attain high tunneling probability for carriers. There are a variety of on-going efforts to fabricate TMOSFETs with Si materials [2-5,8]. The high energy gap - which provides the barrier to tunneling between- source and channel- tends to decrease the tunneling probability and thus reduce

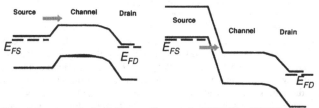

Figure 2: Representative band diagrams of present-day MOSFET and TMOSFET showing different injection mechanisms from source to channel.

the ON current of the device. A preferable approach is to employ a heterojunction between source and channel - that is, employ different materials for these two regions - and to choose the materials so that their band alignment involves a staggered configuration [5-7].

The set of III-V semiconductors provides a rich array of bandgaps and band lineups that can be exploited to make heterojunctions of different characteristics. Figure 3 shows pictorially the band lineups of some familiar binary and ternary III-V materials (although not all are lattice-matched). In order to achieve a high tunneling probability, a choice of materials that involves a small interface bandgap is appropriate, typified by the lineup between InGaAs and GaAlSb. The tunneling probability across such a heterojunction can be dramatically higher than for the case of Si. Figure 4 illustrates, for example, the tunneling current density expected on the basis of

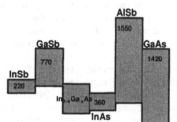

Figure 3: Band lineup of several III-V semiconductors.

4

simple Zener tunneling theory, for tunnel diodes prepared with Si and with a III-V semiconductor choice with interface bandgap (E_{int}) of 0.2eV. The considerable benefit of III-V semiconductor relative to Si is clear.

III-V MOSFET design entails a variety of considerations that differ from the Si case: 1) higher peak velocity and ballistic injection velocity; 2) reduced density of states, leading to lower current levels at a fixed fermi level vs Si ; 3) need for electrostatic confinement to shield the drain potential in small FETs, which is difficult to achieve with the larger wavefunction spread typical of III-Vs ; 4)"source starvation" in ballistic FETs, associated with the difficulty of maintaining thermal equilibrium in the source during ballistic current injection ; 5) oxide issues and interface characteristics. This paper describes numerical simulations of device characteristics,

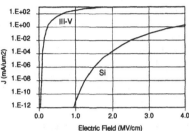

Figure 4: Tunneling current density computed for Si p-n junctrion and for III-V heterojunction with 0.2eV interface bandgap, using simple Zener theory.

and tradeoffs in device design in relation to band offsets, doping levels, and gate geometry.

TMOSFET STRUCTURE

The implementation of TMOSFETs using III-V materials can be done with several geometries such as the one shown schematically in Fig. 5a. The source consists of p-AlGaSb, which injects electrons into the channel of InGaAs (which is doped lightly p type or is undoped, so that the current flow without gate voltage applied is small). The drain is n+ InGaAs or other

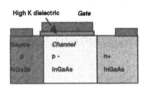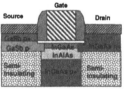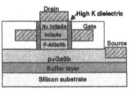

Figure 5: a) Schematic structure of III-V TMOSFET; b) detailed structure with lateral geometry; c) detailed structure in vertical embodiment.

material with potentially large bandgap. Practical implementations of this structure could be have both lateral or vertical geometries, as shown in figure 5b,c. The lateral geometry is fabricated using

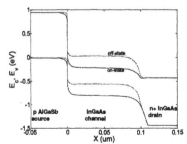

Figure 6: Simulated band diagram of n-channel TMOSFET between source and drain, under different VGS biases.

epitaxial regrowth for the source (and potentially the drain, as shown in the figure). A regrowth process for source/drain regions of III-V MOSFETs has already been demonstrated at UCSB [10]. The vertical structure can be mapped into a pillar or nanowire embodiment in which the gate is wrapped around the entire device.

The band diagram between source and drain along the channel surface for a representative structure is shown in Fig. 6. Under control of the gate voltage, the band lineup between channel and source is varied between the condition where no tunneling can occur (OFF state) and the condition where tunneling is permitted because of the overlap in energy of empty states from the channel conduction band and filled states of the source valence band.

The detailed choice of materials is subject to the constraint of approximately matching lattice constants to avoid misfit dislocation formation. Fortunately there is a rich set of materials including ternary and quaternary III-V compounds from which heterojunction combinations can be selected such that the thin layers needed for TMOSFETs are within the allowed critical thicknesses, or are even lattice-matched (for example, by choosing AlGaAsSb instead of AlGaSb [7]). The material bandgap and band offset estimates used in the design must account for the effects of strain as well as of quantum confinement of the carriers in the thin channel layer.

While figure 6 depicts an n-channel device, it is possible to implement similar structures for p-channel operation with the same material system.

TMOSFET CURRENT-VOLTAGE CHARACTERISTICS

To assess the performance of the TMOSFETs, we have carried out calculations employing 2D solution of current transport and Poisson equations, using the DESSIS software, taking into account direct band-to-band tunneling between source and channel. The tunneling current calculated within the DESSIS program has been shown to be in good accord with the results of a direct calculation based on a two-band model and the WKB approximation [6,7]. Resultant device characteristics are shown in figure 7 (for $V_{DS}=0.3V$). The turn-on characteristics show a sub-threshold slope comfortably below 60 mV/decade, and drain current of 0.5 mA/um of gate width is achieved at $V_{GS}=0.3$ V. The ON/OFF current ratio is 10^4-10^5. These characteristics are very favorable for future highly scaled CMOS circuits. The computed characteristics also exhibit leakage current for negative V_{GS} values, associated with tunneling current at the drain channel junction (which is dependent on the choice of material for the drain).

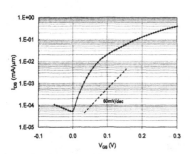

Figure 7: Id-Vgs characteristics of GaAlSb/InGaAs TMOSFETs simulated numerically as described in text.

MATERIALS AND FABICATION CHALLENGES

There are a variety of design issues critical to the performance of TMOSFETs. Well-established limits to the current per unit gate width of III-V MOSFETs are related to the velocity of carriers at the top of the source/channel barrier ("virtual source") and the density of carriers at that position, for a given applied gate voltage. The number of carriers attainable with III-V channels is

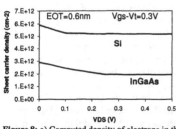 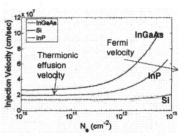

Figure 8: a) Computed density of electrons in the channel of conventional MOSFETs based on Si and on InGaAs at the virtual source, for a gate overdrive of 0.3V, and EOT of 0.6nm. b) Computed average velocity of electrons at the virtual source in conventional MOSFETs, as a function of density of carriers.

limited by the low density of states (DOS) for the conduction band of these materials. For example, figure 8a illustrates the computed channel charge density for Si and for n-In$_{0.53}$Ga$_{0.47}$As channels vs drain voltage, for gate overdrive of 0.3V and equivalent oxide thickness (EOT) of 0.6nm. The factor of x2.5 advantage of Si is apparent, as a result of its higher DOS (in turn related to its higher electron effective mass as well as the degeneracy of the conduction band minima). The low effective mass of InGaAs, however, leads to higher velocity of carriers at the virtual source. At low gate overdrives this advantage follows the thermal velocity (square root of the m* ratio), as expected from Boltzmann statistics. However, at realistically high gate overdrive values, the velocity improvement of III-V semiconductors increases considerably as a result of band-filling and the resultant increase in the fermi velocity of carriers in the degenerate conduction band. The improvement in velocity of x4 over Si leads to an overall increase in the drain current expected for the III-V MOSFET embodiment, according this simple quasi-equilibrium picture. Additional considerations modify this picture, associated with the extent to which the states in the conduction band of the channel can be filled from the source [12]. These effects limit the drain current that can be conducted in traditional MOSFETs, for a given value of gate voltage, to values in the range of 1-2 mA/um for a gate overdrive of 0.3V. The drain current in a TMOSFET is strictly limited by this value, and will be smaller if the tunneling probability is not sufficiently high.

The OFF current of TMOSFETs (for V$_{GS}$=0V and V$_{DS}$=0.3V, for example) is related to carrier generation other than through tunneling at the channel/source junction, as shown in figure 9.

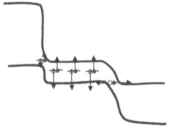

Figure 9: Band diagram of TMOSFET showing schematic carrier generation processes that can contribute to OFF current.

In TMOSFETs based on homojunctions, tunneling at the channel/drain junction also contributes to carrier generation. This component can be suppressed by choosing a material combination featuring an appropriately high interface bandgap at the drain/channel heterojunction. Additionally, electron-hole generation can occur within the channel itself (particularly if the material has a low bandgap) and at the source/channel interface. Since at this last position the interface energy gap is constrained to be low in order to achieve sufficiently high tunneling rates, the interfacial defect density must be minimized.

The implementation of MOSFETs with III-V semiconductors has been hampered for decades by the lack of a suitable oxide or dielectric featuring low interface state densities. Recent results have been very promising; GaAs/InGaAs MOSFETs have been demonstrated employing (Gd,Ga)oxide dielectrics and Al$_2$O$_3$ dielectrics, among others [13-15]. Interface states between channel and gate dielectric have two detrimental effects on TMOSFETs. One effect is to increase the rate of carrier generation at the interface between channel and source, which leads to an increase in leakage current in the OFF state, as described above. Another effect is to electrostatically shield the channel / source interface potential from the effects of the applied gate voltage. Figure 10 illustrates schematically the gate geometry for the TMOSFET. Capacitance C$_{it}$ associated with the interface density of states D$_{it}$ (with C$_{it}$=qD$_{it}$) shunts C$_d$, the geometric capacitance

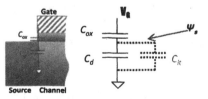

Figure 10: Schematic diagram showing components of input capacitance, and factors that determine the channel potential swing. The effect of interface states is shown.

of the source/channel interface, thus leading to an increase in the value of V$_{GS}$ required to achieve a specific surface channel potential change. Dielectrics with small equivalent oxide thickness values (EOT below about 1 nm) are required for proper operation of TMOSFETs. Smaller EOT values will increase tolerance to presence of interface states. In order to achieve the proper capacitance ratios, it is also important to accurately align the gate with respect to the heterojunction.

An additional materials technology required to allow the TMOSFETs described here to be effective in large scale, high performance CMOS circuits is the deposition of appropriate III-V materials on large Si substrates with low defect density. Vast strides have been made in recent years in the associated growth technology [16].

CONCLUSIONS

TMOSFETs implemented with III-V semiconductors using staggered heterojunction band line-up have the potential to solve the critical power density challenge faced by emerging CMOS technology. TMOSFET operation may well become one of the most compelling reasons to pursue III-V materials for CMOS applications. There remain, however, numerous challenges in material and fabrication technology for the successful implementation of this technology.

ACKNOWLEDGMENTS

The authors would like to thank M. Fischetti, M. Rodwell, P. McIntyre for numerous illuminating discussions of the topics of this paper. Work at UCSD was partially supported by the National Science Foundation.

REFERENCES

[1] Y. Taur, "CMOS design near the limit of scaling", IBM Journal of Research and Development, Volume 46, 2002, pp. 213-222.
[2] Q. Zhang, W, Zhao, and A.Seabaugh, "Low-subthreshold-swing tunnel transistors", IEEE Electron Device Letters, IEEE April 2006 Volume: 27, 297- 300.
[3] NV Girish, R Jhaveri, and JCS Woo, "Tunnel source MOSFET: a novel high performance transistor" - Silicon Nanotechnology Workshop (SNW), Proc. of, 2004.
[4] W. Y. Choi; B.-G. Park; J. D. Lee; T.-J. King Liu, "Tunneling Field-Effect Transistors (TFETs) With Subthreshold Swing (SS) Less Than 60 mV/dec", IEEE EDL Vol. 28, Aug, 2007, 743-745
[5] O. Nayfeh, C. Chleirigh, J. Hennessy, L. Gomez, J. Hoyt and D. Antoniadis, "Design of Tunneling Field-Effect transistors Using Strained-Silicon/Strained Germanium Type-II Staggered Heterojunctions", IEEE Electr. Dev. Letts, Vol. 29, 1074 (2008).
[6] L.Wang, and P. Asbeck, "Design Considerations for Tunneling MOSFETs Based on Staggered Heterojunctions for Ultra-Low-Power Applications", IEEE Nanotechnology Materials and Devices Conference, June 2009, p.196-199.
[7] L. Wang, E.Yu, Y. Taur, and P. Asbeck, "Design of Tunneling Field-Effect Transistors Based on Staggered Heterojunctions for Ultralow-Power Applications", IEEE Electron Device Letters, 2010, Volume 31, p. 431.
[8] Th. Nirschl, et al., "Scaling properties of the tunneling field effect transistor (TFET): Device and circuit" Solid-State Electronics, 50, 2006, 44-51.
[9] J. Appenzeller, Y.-M. Lin, J. Knoch, and Ph. Avouris, "Band-to-Band Tunneling in Carbon Nanotube Field-Effect Transistors", Physics Review Letters, 93, Nov, 2004.
[10] U. Singisetti et al., "InGaAs channel MOSFETs with self-aligned InAs source/drain formed by MEE regrowth", IEEE Electr. Dev. Lett, Nov. 2009.
[11] R. Chau, S. Datta, and A. Majumdar, "Opportunities and Challenges of III-V Nanoelectronics for Future High-Speed, Low-Power Logic Applications", Tech. Dig., 2005 Compound Semiconductor IC Symposium, p.17.
[12] Fischetti, M. V.; Wang, L.; Yu, B.; Sachs, C.; Asbeck, P. M.; Taur, Y.; Rodwell, M.,

"Simulation of Electron Transport in High-Mobility MOSFETs: Density of States Bottleneck and Source Starvation", Tech. Dig. 2007 IEDM, pp. 109-112.

[13]   M. Passlack, et al., "High Mobility Ill-V MOSFETs For RF and Digital Applications", Tech. Dig. 2007 IEDM, p. 621.

[14]   Y. Xuan, Y.Q. Wu, T. Shen, T. Yang and P.D. Ye, "High Performance submicron inversion-type enhancement-mode InGaAs MOSFETs with ALD Al2O3, HfO2, and HfAlO as gate dielectrics", Tech. Dig. 2007 IEDM, p. 637-640.

[15]   E.J.Kim et al., "Atomically abrupt and unpinned $Al_2O_3/In_{0.53}Ga_{0.47}As$ interfaces: Experiment and simulation" J. Appl. Phys. 106, 124508 (2009).

[16]   M. Hudait et al., "Heterogeneous Integration of Enhancement Mode In0.7Ga0.3As Quantum Well Transistor on Silicon Substrate using Thin Composite Buffer Architecture for High-Speed and Low-Voltage (0.5V) Logic Applications", Tech. Dig. 2007 IEDM, p. 625-628.

Mater. Res. Soc. Symp. Proc. Vol. 1252 © 2010 Materials Research Society   1252-I02-09

Tuneable CMOS and Current Mirror Circuit with Double-Gate Screen Grid Field Effect Transistors

Y. Shadrokh[1], K. Fobelets[1], and J. E. Velazquez-Perez[2]
[1]Department of Electrical and Electronic Engineering, Imperial College London, Exhibition Road, London SW7 2BT, UK
[2]Departmento de Fisíca Aplicada, Universidad de Salamanca, Edificio Trilinüe, Pza de la Merced s/n, E-37008 Salamanca, Spain

ABSTRACT

The multiple-gate aspect of the Screen Grid Field Effect Transistor (SGrFET) increases functionality and reduces component count of circuits. An independently-driven gate SGrFET is used to control the switching voltage as well as the gain factor of an inverter. The multi-gate configuration of the SGrFET allows a decrease in output conductance without an increase of transistors count. This leads to a reduction in fabrication complexity, chip area and parasitics. In addition, a simple SGrFETs-based current mirror circuit is proposed with gain factor control.

INTRODUCTION

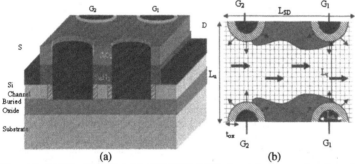

(a)   (b)

Figure 1. (a) 3D schematic of the SGrFET. (b) Horizontal cross section through the channel. Bold arrows indicate the direction of carrier flow, thin arrows the direction of channel depletion.

The unique gate configuration and planar character of the Screen Grid Field Effect Transistor (SGrFET) makes it suitable for digital applications with reduced circuit complexity, reduced gate-drain capacitance, improved sub-threshold slope and increased switching speed. The geometry of a unit cell SGrFET [1] is given in Fig. 1 with definition of the geometrical parameters: source-drain distance: L_{SD}, unit cell width L_u, channel width L_c, gate cylinder diameter, L_0, oxide thickness t_{ox}. The gate consists of multiple cylindrical cavities (fingers) with oxide sidewalls and a poly-Si/metal filling. These fingers are standing perpendicular to the current flow. Source/drain are highly doped and have the same width as the device, reducing contacting problems. For optimum performance the channel doping is low to un-doped in order to preserve high mobility values and is of the same doping type as the contact regions. The device operation is essentially MESFET-alike. The role of the second row of fingers (at drain) is to control short channel effects[1]. In [1] the advantages of logic SGrFET circuits were presented and compared to finFET [2] circuits. In this manuscript we optimise the inverter circuit using the unique geometrical character of the SGrFET. All the simulations are carried out using the 2D TCAD device simulator, Medici^{TM} [3]. In section 1 the rise/fall times of the CMOS inverter is improved while simultaneously increasing the pMOS off and on current. In section 2 an inverter with tuneable transfer characteristics is presented and section 3 a current mirror with tuneable gain.

DEVICE STRUCTURE AND MODELING

The geometrical parameters for the simulations are: L_o = 50nm, t_{ox} = 2nm, L_{SD} = 140nm and L_u = 104nm. The hydrodynamic models are used for the sub-micron dimensions of the device. The field and concentration dependent mobility models are used.

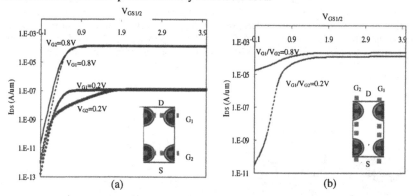

Figure 2. Transfer characteristics of independently-gated SGrFETs (a) gates cylinders connected across the device width (CAW). (b) Gate cylinders connected along the device length (CAL).

Two independently-connected gate configurations and their transfer characteristics are given in fig. 2. Their performance parameters are given in Table 1 for V_{DS} = 1V, and V_{G1} = 0.2V or 0.8V while -0.1V < V_{G2} < 4V. Comparison shows a better performance of the CAW configuration. This is due to better gate control by the CAW connected gate cylinders. Moreover, in CAW, best performance is obtained when the drain side gate cylinders are kept constant while the source side gate cylinders vary (Table I) as a consequence of the shielding action of the drain side gates.

Table I. Performance parameters of the CAW and CAL configuration for V_{DS} = 1V and V_{G1} = 0.2V or 0.8V and -0.1V < V_{G2} < 4V. V_{th}: threshold voltage, S: sub-threshold slope.

	I_{OFF} (pA/μm^2)		I_{ON} (A/μm^2)		V_{th} (V)		S (mV/dec)	
V_{G1} (V)	0.2	0.8	0.2	0.8	0.2	0.8	0.2	0.8
CAW	3.42	386	110	208	0.42	-0.22	76	76
CAL	77	20.7 10^6	0.11	125	0.19	0.34	85	823

In CAW V_{G2} switches the device ON and OFF while the ON current is determined by V_{G1}. However CAL loses its performance in ON and OFF states depending on the voltage on G1, because for $V_{G1} = 0.2V$ one side of the FET is continuously inactive, resulting in a reduction in the gate control. Similarly, for $V_{G1} = 0.8V$ one side of the FET is always active resulting in an increase in the OFF current. ON currents are also higher in the CAW configuration.

TUNABLE CMOS VOLTAGE TRANSFER CHARACTERISTICS

Two inverter circuits (fig. 3), based on the CAW and CAL configuration are simulated. One of the gates is connected to the input and a constant bias is applied to the other gate connection.

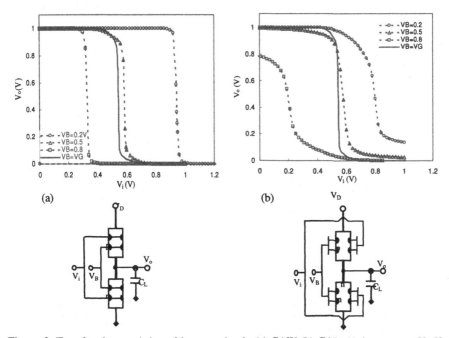

Figure 3: Transfer characteristic and inverter circuit. (a) CAW (b) CAL. V_B is constant, $V_i = V_G$ input, V_o output. C_L: load capacitance of the following stage. $V_B = V_G$ all gates change together.

When keeping V_B constant and varying V_i between 0V and 1V, an inverter output is obtained in which the switching voltage of the output is controlled by the value of V_B (fig. 3). If $V_B > V_{th-n}$ and $V_G < V_{th-n}$ only half of the n- and p-type FETs are functioning. Once $V_G > V_{th-n}$, the pull down FET turns ON completely while only half of the pull-up FET conducts. This results in the output characteristic shifting towards the left of the single gate inverter ($V_B = V_G$). The magnitude of the shift in output voltage switching point as a function of V_B is given in Table II. This tuneable inverter is suitable for controlling both voltage and current gain. Note that a similar shift in output voltage transfer characteristics can be obtained by using minimum four

MOSFETs[4]. Performance of the inverter can be optimised by changing the size and position of the gate cylinders and by applying techniques as presented in[5].

Noise margins and switching times based on techniques presented in [6] are given in Table III.

Table II. Shift of the output switching voltage as a function of the constant bias voltage V_B.

V_B (V)	CAW - $V_{switching}$ (V)	CAL - $V_{switching}$(V)
0.2	0.32	0.19
0.5	0.57	0.56
0.8	0.94	0.76

Table III. Noise Margin (NM) for the independently inverter circuits in fig. 3 at V_B=0.5V.

	CAW inverter	CAL inverter	V_B=V_G
NM_L(V)	0.50	0.38	0.41
NM_H(V)	0.31	0.29	0.40
Rise time (ps)	13	17.1	18.9
Fall time (ps)	30.2	14.2	10.1

Complete switching is achieved with the CAW configuration with larger noise margins compared to the CAL configuration. The CAL configuration shows no full OFF and ON states. Unlike the excellent ON and OFF switching times obtained in the classical SGrFET CMOS (V_B=V_G), the independently-driven gate SGrFETs in the CAW configuration shows significantly increased fall times, due to the larger gate-drain capacitance[7]. Although the CAL FET shows improved switching speed, this is limited to a certain gate voltage range. The CAW inverter shows better switching times when $V_B < V_{DD}/2$. In order to improve the fall time, the gate geometry of the CAW FET can be adapted to reduce the gate-drain capacitance. This can be done by using the techniques presented in [8] and [9].

GAIN FACTOR IN SGrFET CURRENT MIRROR

In ref. [10]a tuneable current mirror circuit with double gate MOSFETs was proposed, using dual metal gate processing. The same is possible with SGrFETs but using a single gate metal with a work function of 4.8eV. The threshold voltage of the single unit cell n-type devices is V_{th} = 0.4V at V_{DS} = 1V. Fig. 4 shows two current mirror circuits one based on single gate bias and the other on a multi-gate bias and their output characteristics. Tuneable current mirrors that use the SGrFET with multiple unit cells have the advantage that the current gain (β) can be increased by an increase in the number of unit cells (N) (see Table IV(a)). Thus β increases by a factor of ~1.5 times for each added unit cell, due to an increase in current level with N [1]. In addition, the output current can be tuned via the double gate approach. In our single gate circuit a multi-cell SGrFET is required to obtain $\beta > 1$. β can be changed by V_B as given in Table IV(b).

Table IV. (a) Gain factor β and minimum output transconductance g_{dmin} as a function of number of unit cells N. (b) β as a function of V_B at $V_{DD2} = 1$V and $V_G = 1$V.

N	β (dB)	L_o=25nm	L_o=30nm
		g_{dmin} (S/m²)	g_{dmin} (S/m²)
1	12.5	0.464	0.199
2	22.5	0.926	0.397
3	30	1.45	0.556

(a)

V_B (V)	β (dB)
0.6	10.3
0.7	15.4
0.8	20.6
0.9	25.6
1	30

(b)

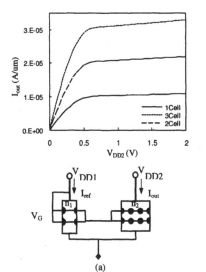

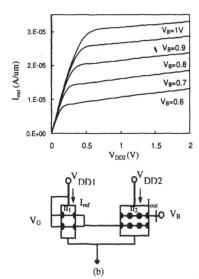

(a)   (b)

Figure 4. Output Current versus output voltage (a) as a function of the number of unit cells and (b) as a function of V_B. I_{ref}=10.4 μA/μm².

The low output transconductance in current mirrors using the double-gate MOSFET was fixed using modified cascade current mirror circuits with 4 devices[10]. A low output admittance of the SGrFET current mirror can be obtained by an increase in gate diameter without increasing the number of devices. If the gate diameter, L_o is increased to 30 nm without changing the other dimensions then the output conductance improves as given in Table IV(a). This improvement is due to the reduction of the channel width.

CONCLUSION

Two different independently-driven double-gate SGrFETs are investigated. It is shown that the double gate SGrFET with gate cylinders connected across the width of the device (CAW) shows better performance than the SGrFET with gate cylinders connected across the length (CAL). The independent-double-gate SGrFETs in a CMOS circuit can shift the switching voltage of the transfer characteristics over a range of 0.62V via a constant voltage applied to one set of the gate cylinders. The CAW device structure shows the best performance in terms of complete ON and OFF states and noise margin. Rise and fall times were smaller for the CAL FET for $V_G < V_{DD}/2$ otherwise the CAW device configuration shows faster switching.

The current mirror designed using only two CAW SGrFETs gives current gain control using an independent-double-gate SGrFETs with one constant gate control voltage or via an increase in the number of the unit cells. The output resistance of the current mirror can be reduced by an increase in gate cylinder diameter without an increase in the number of devices in the mirror current.

ACKNOWLEDGEMENT

This research was financially supported by the EPSRC under grant number EP/E023150/1. Discussions with Dr. E. Rodriguez-Villegas are gratefully acknowledged.

REFERENCES

[1]   K. Fobelets, P. W. Ding, and J. E. Velazquez-Perez, "A novel 3D Gate Field Effect Transistor Screen-Grid FET Device-Concept and Modelling," *Solid State Electron*, vol. 51 (5), pp. 749-759, 2007.

[2]   Y. Shadrokh, K. Fobelets, and J. E. Velazquez-Perez, "Comparison of the multi-gate functionality of screen-grid field effect transistors with finFETs," *Semicond. Sci. Technol*, vol. 23, pp. 9, 2008.

[3]   "Taurus-Medici User Guide (Version W-2004.09), S.I., Mountain View, CA."

[4]   D. A. Hodges, H. G. Jackson, and R. A. Saleh, *Analysis and Design of Digital Integrated Circuits: In Deep Submicron Technology*: McGraw-Hill Professional, 2004.

[5]   S. Kaya, H. F. A. Hamed, and J. A. Starzyk, "Low-Power Tunable Analog Circuit Blocks Based on Nanoscale Double-Gate MOSFETs," *IEEE Transactions on Circuit and Systems*, vol. 54, pp. 571-575, JULY 2007.

[6]   J. S. Yuan and L. Yang, "Teaching Digital Noise and Noise Margin Issues in Engineering Education," *IEEE Transactions on Education*, vol. 48, pp. 162-168, 2005.

[7]   B. Andreev, E. L. Titlebaum, and E. G. Friedman, "Sizing CMOS Inverter with Miller Effect and Threshold voltage Variations," *World Science*, vol. 15, pp. 437-454, 2006.

[8]   M. Masahara, R. Surdeanu, L. Witters, G. Doornbos, V. H. Nguyen, G. V. d. bosch, C. Vrancken, K. Devriendt, F. Neuilly, E. Kunnen, M. Jurczak, and S. Biesemans, "Demonstration of Asymmetric Gate Oxide Thickness 4-Terminal FinFETs," *IEEE Electron Device Letters*, vol. 28, pp. 217-219, 2007.

[9]   Y. Shadrokh, K. Fobelets, and J. E. Velazquez-Perez, "Optimizing the Screen-Grid Field Effect Transistor for high drive current and low Miller capacitance," *2009 Spring MRS Conference*, 2009.

[10]   H. F. A. Hamed, S. Kaya, and J. A. Starzyk, "Use of nano-scale double-gate MOSFETs in low-power tunable current mode analog circuits," *Analog Integrated Circuits and Signal Processing*, vol. 54, pp. 211-217, 2008.

Mater. Res. Soc. Symp. Proc. Vol. 1252 © 2010 Materials Research Society   1252-I02-10

A new SiGeC Vertical MOSFET: Single-Device CMOS (SD-CMOS)

Carlos J. R. P. Augusto and Lynn Forester
Quantum Semiconductor LLC, 4320 Stevens Creek Blvd., Suite 212,
San Jose, CA 95129, U.S.A.

ABSTRACT

A new type of silicon-based Vertical MOSFET concept is presented, Single-Device CMOS (SD-CMOS), in which the same structure can be operated as NFET or as PFET, depending on the biasing conditions [1,2]. SD-CMOS offers new possibilities for simpler CMOS integration schemes; one of them requiring only 4 masks for the "Front-End"; with less cost to manufacture than any integration scheme requiring the fabrication of two devices with opposite doping polarities. Numerical simulations with a commercial device simulator [3] confirm the validity of the concept and demonstrate its feasibility for scaling to 10nm channel lengths.

INTRODUCTION

V-MOSFETs made by epitaxial growth can easily have channel lengths smaller than 10nm (L_{Ch}<10nm), but DIBL (Drain Induced Barrier Lowering) and OFF-state current have remained as major device physics obstacles, and the threshold voltage (V_T) cannot be controlled reliably by "intermediate" doping levels, in part due to the significant statistical dopant fluctuations in such small active volumes.

For V-CMOS, in which both NMOS and PMOS are epitaxially grown, a CMOS process integration architecture has not yet been identified or widely accepted.

SD-CMOS overcomes the device physics shortcomings of the previous types of V-MOSFETs, and the fabrication of a single set of device layers (Source/Channel/Drain) radically simplifies the Front-End process flow for complementary V-MOSFETs.

THEORY

A schematic cross section of Single-Device CMOS (SD-CMOS) is shown in **FIG. 1**, with the corresponding energy band diagram, shown in **FIG. 2**, for zero bias condition.

A few key differences with respect to other V-MOSFETs should be readily noticed: (1) the source is a non-doped semiconductor region with a very narrow band-gap; (2) the channel is a non-doped semiconductor with a wider band-gap, having nearly identical conduction and valence band offsets with the source region; and (3) the drain is a metal with a work-function (W_F) near the mid-gap level of the channel and source materials. The source region is contacted by a metal with a W_F corresponding to its mid-gap level, and thus to the mid-gap level of the channel. The source region, although non-doped, has a band-gap that is narrow enough (e.g., 0.1eV) to have a high intrinsic carrier concentration, a negligible barrier height (<3KT) at room temperature and above, for electrons and holes. An energy band diagram, from source to drain, along a vertical cut made near the interface with a gate, will be identical to that made at the mid-distance point between gates, if the gate electrode has a mid-gap W_F, otherwise the bands in the source and in the channel regions will bend accordingly.

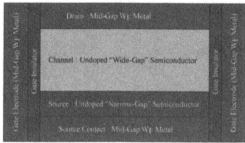
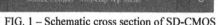
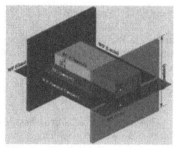

FIG. 1 – Schematic cross section of SD-CMOS   FIG. 2 – 3D view of band diagram.

This type of band structure allows an arbitrary distance between opposing gates without affecting the potential barriers at the source and the channel regions, thereby making the threshold voltage (V_T) independent of that distance and independent of patterning (lithography + etching). The resulting potential barriers are <u>symmetric</u> from source to drain, with respect to the midgap energy level (the "mirror" line). On the other hand, the conduction and valence band (CB and VB) edges are required to be <u>asymmetric</u> with respect to the source/channel and channel/drain interfaces, and therefore the source and drain regions are not interchangeable.

At the source/channel interface, the CB (VB) offset between source and channel sets the barrier height for electrons (holes) in the OFF condition for NMOS (PMOS). Applying a voltage at the gate leads to the accumulation of electrons (holes) at the source/channel interface, thereby pushing the Quasi Fermi-Level for electrons (holes) in the source above (below) the CB (VB) edge of the channel for the ON condition. The field effect of the gate modulates the relative position of the CB or VB edges with respect to the Quasi Fermi-Levels in the source region. By defining an "Effective Barrier Height" (EBH) as the distance (in energy) between the CB or VB edges of the band in the channel and the Quasi Fermi-Levels in the source, it is possible to see that this EBH can be controlled by the field effect of the gate. Such effect cannot be obtained in the case where the source/channel interface is a Schottky junction with a "fixed" height, and in which the gate can only control tunneling probability through the Schottky barrier. In SD-CMOS, due to the symmetry of the potential profile for electrons and holes from source to drain, the exact same structure operates as a N-MOSFET when $V_{DS}>0$, $V_{GS}>0$, and operates as a P-MOSFET when $V_{DS}<0$, $V_{GS}<0$. The Schottky barrier at the channel/drain interface prevents current injection from drain into channel, and insures that a NMOS in the "OFF" state is not, simultaneously, a PMOS in the "ON" state, and/or vice-versa.

The band diagrams suggest that SD-CMOS should overcome some of the drawbacks of conventional ultra-small Si MOSFETs, such as DIBL and "source starvation", as well as enabling and maintaining performance scaling as a function of channel length, for 10nm channel lengths and below [4].

Implementation of SD-CMOS with Silicon-based materials

SD-CMOS can be implemented with $Si_{1-x}Ge_x$, and/or $Si_{1-y}C_y$ and/or $Si_{1-x-y}Ge_xC_y$, strained to a Si substrate. The narrow-gap source region is more challenging than the channel, but can be implemented with strained MonoLayer (ML) SuperLattices (SLs). These can generate a synthetic band structure along the direction of epitaxial growth, which is also the direction of current flow in V-MOSFETs, including SD-CMOS. Key parameters such as band-gap type and magnitude can be strongly affected by varying the period of the SL and the composition of its constituents. The requirements for the source region of SD-CMOS can be met with an SL of alternating "m" MLs of a $(Si_{1-y}C_y)$ random alloy with "n" MLs of pure Ge (or $Si_{1-x}Ge_x$, with high Ge content), which generates a type-II band alignment. The VB of the pure Ge layer is pushed upwards by at least 0.72eV [5], and the CB edge of the $(Si_{1-y}C_y)$ random alloy is pushed downwards, which for films on (100) Si substrates leads to $\Delta E_C^{(\Delta2)} = -y \cdot 4.6eV$ [6]. For sufficiently high carbon content, the CB edge of the $(Si_{1-y}C_y)$ random alloy crosses the top of the VB of the Ge layer, thereby forming a "broken-gap" semiconductor, or semi-metal.

The band offsets of the source $(Si_{1-y}C_y)_m$-$(Si_{1-x}Ge_x)_n$ SL minibands, with respect to bulk Si, will not be symmetric, and therefore the channel material should not be pure Si, but rather: (1) a $Si_{1-x-y}Ge_xC_y$, random alloy whose composition is adjusted to produce symmetric band off-sets with respect to the source region; or (2) a $(Si_{1-y}C_y)_m$-$(Si_{1-x}Ge_x)_n$ SL with a different period and/or C and/or Ge content, in order to produce the desired symmetric offsets between the mini-bands in the source and the minibands in the channel. Similarly, the choice of metals for the source contact metal, drain, and gate electrodes, is adjusted to produce the desired electrostatic configuration, i.e., W_F aligned with the mid-gap level of the source and channel.

Different gate electrode W_F can be selected to adjust V_T without changing the physical picture in the region between opposing gates. With a "gate-last" approach, it is straightforward to fabricate gates with dual metal electrodes having W_F corresponding to N+ and P+ poly-Si, there-by lowering the V_T for NMOS and PMOS. Although V_T can be changed by changing EBH, this affects the OFF-current across the area between the opposing MIS gates, not just near the MIS gate interfaces. Ideally, EBH should be such that OFF-current is minimized and balanced for both NMOS and PMOS. Also the distance between opposing MIS gates could also be reduced to generate quantization subbands, which will increase the EBH in the region between the opposing gates, consequently decreasing the OFF-current, and increasing V_T.

DISCUSSION
Numerical Simulations of SD-CMOS

Numerical simulations, using energy balance models, were carried out to validate the theoretical concept of a MOSFET that can operate as NMOS or PMOS, depending only on the applied voltage. The simulations were performed with the following device parameters: thick-ness of source region = 4 nm; thickness of the channel region 10nm; gate insulator (SiO₂) thick-ness 0.5nm; and distance between opposing gates = 100nm. The simulator used does not have the ability to model SL minibands, but the principles of operation can be verified for a source region with a band-gap set to 0.1eV at room temperature, while all other variables, such as effec-tive mass, dielectric constant, etc., were those of silicon, which results in an intrinsic carrier con-centration of ~4.6E18 cm⁻³. The CB and VB offsets were set to be identical at the source/channel interface. To demonstrate that the same structure can operate as a NMOS or PMOS, the W_F of the gate electrode was chosen to be equal to that of the mid-gap level of the channel (Si), i.e., around 4.65eV. The band alignments are shown in **FIG. 3**. In simulations of operation as NMOS, V_{DS} was ramped from 0 to +1V, followed by the ramping of the V_{GS} from 0V to +1V. In simulations of operation as PMOS, V_{DS} was ramped from 0 to -1V, followed by the ramping of the V_{GS} from 0V to -1V. Additional simulations were performed for gate electrodes with the W_F values of 4.05eV (equivalent to N+-poly) and 5.2eV (equivalent to P+-poly). As expected, the V_T for NMOS and PMOS were lowered. The results of the simulations are shown in **FIG. 4**.

Referring to **FIG. 4**, it should be noted that when SD-CMOS operates as NMOS, the hole-current (hCurrent) is several orders of magnitude smaller than the electron current (eCurrent), which is a result of the much more efficient forward injection of electrons across the source/channel interface, than the reverse injection of holes across the drain/channel interface. The situation is analogous and reversed for SD-CMOS operating as a PMOS. The band diagrams for SD-CMOS as NMOS are shown in **FIGs. 5A, 5B, and 5C**, and the band diagrams for SD-CMOS as PMOS are shown in **FIGs. 6A, 6B, and 6C**. In **FIGs. 4, 5A-5C, and 6A-6C**, the source/channel interface is positioned at the zero point in the x-axis.

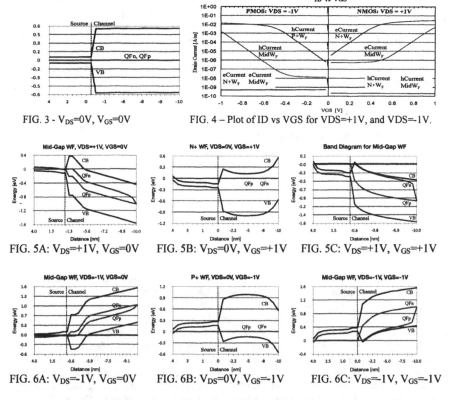

FIG. 3 - V_{DS}=0V, V_{GS}=0V   FIG. 4 – Plot of ID vs VGS for VDS=+1V, and VDS=-1V.

FIG. 5A: V_{DS}=+1V, V_{GS}=0V   FIG. 5B: V_{DS}=0V, V_{GS}=+1V   FIG. 5C: V_{DS}=+1V, V_{GS}=+1V

FIG. 6A: V_{DS}=-1V, V_{GS}=0V   FIG. 6B: V_{DS}=0V, V_{GS}=-1V   FIG. 6C: V_{DS}=-1V, V_{GS}=-1V

FIGs. 5A and 6A show that a significant source/channel EBH still exists even for an electric field in the channel of 1V/10nm, that is, twice the breakdown value for bulk Si (0.5MV/cm). DIBL calculated as the decrease in EBH between V_{DS}=0 and V_{DS}=1V, is about 189mV/V for this device with L_{ch}=10nm. Also, **FIGs. 5C and 6C**, show that the gate is able to change the carrier concentration in the narrow gap source (n_i=~4.6E18cm^{-3}) to the extent that the modulation of the EBH is sufficient to reach I_{ON}/I_{OFF}=~1E4.

In all simulations, gate interface states and gate leakage currents are not included since they are not required to prove the ability of SD-CMOS to operate as NMOS or PMOS. These band diagrams indicate that the length, that is, the film thickness, of source and channel regions can be further reduced. It should be highlighted that for L_{Ch}=10nm or less, transport will be ballistic at room temperature, and that in such a regime, device performance increases for heavier masses, thus favoring silicon over high-mobility and small-mass III/V materials, and that the ultimate switching speed for such devices is on the order of 30 THz [7].

Linear and log plots of I_D vs. V_{DS} are shown, respectively in **FIGs. 7A & 7B** for NMOS with N+W_F gates, and in **FIG. 8A & 8B** for PMOS with P+W_F gates. Linear and log plots of I_D vs. V_{GS} are shown, respectively in **FIGs. 7C & 7D** for NMOS with N+W_F gates, and in **FIG. 8C & 8D** for PMOS with P+W_F gates. The linear plots of I_D vs. V_{DS} show that linear operation starts only when V_{DS} approaches the Schottky barrier height. From these simulations it is possible to extrapolate a Subthreshold Slope (SS) of about 120mV/decade, which is a fairly large value, probably due to certain models used in the simulations, such as the silicon-like carrier masses and density of states in the source region, as well as the non-ballistic transport in the 10nm long channel, all of which decrease the drive current (I_D) for the same applied voltage (V_{GS}&V_{DS}).

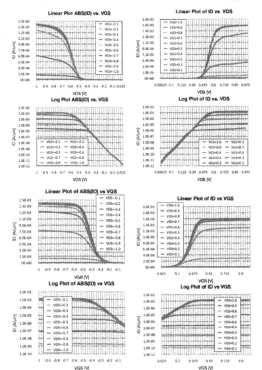

FIG. 7A (PMOS) & 8A (NMOS)
Linear Plots of I_D-V_{DS}.

FIG. 7B (PMOS) & 8B (NMOS)
Log Plots of I_D-V_{DS}.

FIG. 7C (PMOS) & 8C (NMOS)
Linear Plots of I_G-V_{GS}.

FIG. 7D (PMOS) & 8D (NMOS)
Log Plots of I_D-V_{GS}.

Front-End Fabrication Flow using only 4 Masks

An exemplary fabrication flow is described in **FIGs. 9A-9F**, showing an "Inner Gate" layout, which has several advantages in terms of density over the better known "Surrounding Gate", since it takes advantage of the built-in parallelism of V-MOSFETs, in which several gates can share the same source and drain layers and contacts. This is particularly useful for "wide bit" memory cells and NOR logic gates. In any case, both types of gates are made with the same mask. The fabrication of gate stacks on vertically-etched surfaces is no longer seen as an obstacle for V-MOSFETs since it is also required for certain non-planar horizontal devices, such as Fin-FETs. SD-CMOS can be implemented with $Si_{1-y}C_y$, $Si_{1-x}Ge_x$, and Ge layers/materials, well-known and widely accepted materials, and is thus at a considerable advantage over devices incorporating III/V materials for which it is not yet certain if good quality and reliable passivation and ohmic contacts can be developed.

FIG. 9A – Epitaxial deposition of Source/ Channel layers, and deposition of metal Drain, followed by bonding to an insulator substrate.

FIG. 9B – Removal of the substrate on which the epitaxial layers were formed, followed by deposition of the metal Source Contact layer, and deposition of thin SiO_2 and thick Si_3N_4.

FIG. 9C – Mask 1: "Field Isolation", by trench etching through device layers, and filling with SiO_2, followed by CMP planarization stopping on thick Si_3N_4.

FIG. 9D – Mask 2: "Gate Stack", by trench etching through device layers, and filling with gate insulator + gate electrode, followed by CMP planarization stopping on thick Si_3N_4.

FIG. 9E – Mask 3: "Contact to Source", by trench etching through thick Si_3N_4 and thin SiO_2, stopping on source metal, filling with metal, followed by planarization stopping on thick Si_3N_4.

FIG. 9F – Mask 4: "Contact to Drain", by trench etching through device layers, stopping on Metal Drain layer, followed by formation of spacers, filling with metal, and planarization.

Impact of SD-CMOS on Circuit Design

It follows from the device structure of SD-CMOS that source and drain are not interchangeable, and this has immediate repercussions, for example, when using SD-CMOS as a pass-transistor, a fairly simple solution is to replace a single conventional pass transistor with two SD-CMOS devices, such that each of the two nodes is connected to the source of one SD-CMOS device, thereby allowing current to flow in both directions. Therefore, SD-CMOS should be able to transparently replace any conventional NMOS or PMOS, without any modification to circuit design methodologies (apart from adjustments for different V_{DD} and V_T).

Recent reports [8,9,10] on circuit design with ambipolar devices indicate considerable advantages over conventional CMOS circuitry with respect to overall performance, power dissipation, functionality per unit of area, etc., provided that their unique characteristics are fully utilized. Unlike other ambipolar devices, SD-CMOS can operate as NMOS or PMOS depending only on the applied voltage, without requiring control signals (e.g., a "control gate").

There are two distinct scenarios for the incorporation of SD-CMOS into circuitry:

1) Operation as NMOS and PMOS is not dynamically reconfigurable.

In this case some devices could be optimized, preferably just through a different gate electrode W_F, to operate as NMOS or as PMOS. The technologies developed for the "gate last" approach to the fabrication of gate stacks for standard CMOS at 45nm and below, should have a thermal budget low enough to enable the fabrication of multiple gate stacks, thereby making this option as manufacturable as the version with a single gate electrode.

2) Operation as NMOS and PMOS is dynamically reconfigurable.

In this case, the same device should be able to operate as NMOS and PMOS with fairly symmetric electrical characteristics, namely V_T, OFF-current, ON-current, etc., and the potential barriers built during the epitaxial growth process should be approximately the same for electrons and holes, and the gate electrode should have a mid-gap W_F. For the operating voltage possible with such short channels, a gate with mid-gap W_F may not lead to a complete turn-on of the device, but this should not be seen as a technical barrier, since one of the solutions being pursued to overcome the power density problem in advanced CMOS, is to design circuits to be operated in the subthreshold regime [11].

CONCLUSIONS

A new Vertical MOSFET device, SD-CMOS, has been introduced. Simulations show that a single device structure can operate as NMOS or PMOS, depending only on the applied voltage, and is thus reconfigurable by changing the potential at the source. The channel length and threshold voltage are defined by an epitaxial heterojunction, and do not depend on lithography. Sophisticated band-structure engineering is possible along the direction of current flow, thereby enabling the improvement of electron and hole transport characteristics. The heterojunction at the source/channel interface enables 10nm channel lengths without DIBL, and a consistent/reproducible threshold voltage across the entire substrate, not just within a die.

A new process flow has also been introduced in which the "Front-End" of CMOS fabrication requires only 4 masking layers. With an "Inner-Gate" layout scheme, Source & Drain can be shared by several gates in parallel, leading to higher density of integration of functional blocks, such as NOR gates and RAM cells. The proposed materials and fabrication technology leverage existing CMOS process technologies and manufacturing infrastructure.

Transparent replacement of conventional CMOS devices should be possible without major impact on design methodologies, while the full utilization of the new flexibility of ambipolar devices should lead to extremely important gains in performance, power dissipation, area, etc.

From several perspectives, including device physics, fabrication simplicity and cost, utilization of existing CMOS manufacturing infrastructure and equipment, and immunity against critical process variations in sub-22nm CMOS generations, SD-CMOS presents a compelling solution for the "End of Roadmap" CMOS technology.

REFERENCES

1. Carlos J. R. P. Augusto, US Patent No. 6 674 099, (6 January 2004).
2. Carlos J. R. P. Augusto, US Patent No. 7 023 030 (4 April 2006).
3. DESSIS 2D device simulator, ISE Integrated Systems Engineering AG, TCAD release 10.0.
4. M.V. Fischetti, S. Jin, T.-W. Tang, P. Asbeck, Y. Taur, S. E. Laux, N. Sano, IEEE 13th International Workshop on Computational Electronics, 2009, IWCE '09; DOI: 10.1109/IWCE.2009.5091145.
5. J. C. Sturm, H. Manoharan, L. C. Lenchyshyn, M. L. W. Thewalt, N. L. Rowell, J.-P. Noël, D. C. Houghton, Phys. Rev. Lett. 66, pp. 1362-1365, (1991); DOI: 10.1103/PhysRevLett.66.1362.
6. K. Eberl, K. Brunner and W. Winter, Thin Solid Films, 294 (1-2), pp. 98-104 (1997); DOI: 10.1016/S0040-6090(96)09269-3.
7. P. M. Solomon and S. E. Laux, IEDM Tech. Dig., 2001, pp. 95–98; DOI: 10.1109/IEDM.2001.979425.
8. I. O'Connor, J. Liu, F. Gaffiot, F. Prégaldiny, C. Lallement, C. Maneux, J. Goguet, S. Frégonèse, T. Zimmer, L. Anghel, T.-T. Dang, R. Leveugle, "CNTFET Modeling and Reconfigurable Logic-Circuit Design", IEEE Trans. Circ. Syst.—I, Vol. 54, No. 11, Nov. 2007, pp. 2365-2379; DOI: 10.1109/TCSI.2007.907835.
9. W. J. Yu, U. J. Kim, B. R. Kang, I. H. Lee, E.-H. Lee, Y. H. Lee, "Adaptive Logic Circuits with Doping-Free Ambipolar Carbon Nanotube Transistors", NANO LETTERS, 2009 Vol. 9, No. 4, 1401-1405; DOI: 10.1021/nl803066v.
10. K. Jabeur, D. Navarro, I. O'Connor, P. E. Gaillardon, M. H. B. Jamaa, F. Clermidy, "Reducing transistor count in clocked standard cells with ambipolar double-gate FETs", Nanoscale Architectures (NANOARCH), 2010 IEEE/ACM International Symposium on, 17-18 June 2010, pp. 47-52; DOI:10.1109/NANOARCH.2010.5510928.
11. See for example Proceedings of the IEEE, Vol. 98, No. 2, February 2010, "Special Issue on Circuit Technology for ULP", with an introduction by R. H. Reuss, M. Fritze; DOI:10.1109/JPROC.2009.2037210.

Ge MOSFET

Mater. Res. Soc. Symp. Proc. Vol. 1252 © 2010 Materials Research Society   1252-I03-01

Epitaxial Dy$_2$O$_3$ Thin Films Grown on Ge(100) Substrates by Molecular Beam Epitaxy

Md. Nurul Kabir Bhuiyan[*1], Mariela Menghini[1], Christel Dieker[2], Jin Won Seo[3], Jean-Pierre Locquet[1], Roumen Vitchev[4] and Chiara Marchiori[5]
[1]Department of Physics and Astronomy, Katholieke Universiteit Leuven, Celestijnenlaan 200D, B-3001, Leuven, Belgium
[2]Mikrostrukturanalytik, Christian-Albrechts Universität zu Kiel, Kaiserstrasse 2, D-24143 Kiel, Germany
[3]Department of Metallurgy and Materials Engineering, Katholieke Universiteit Leuven, Kasteelpark Arenberg 44, B-3001, Leuven, Belgium
[4]VITO Materials, Flemish Institute for Technological Research, Boeretang 200, B-2400 Mol, Belgium
[5]IBM Research GmbH, Zurich Research Laboratory, Saeumerstrasse 4, 8803, Rueschlikon, Switzerland

ABSTRACT

Dysprosium oxide (Dy$_2$O$_3$) films are grown epitaxially on high mobility Ge(100) substrates by molecular beam epitaxy system. Reflection high energy electron diffraction patterns and X-ray diffraction spectra show that single crystalline cubic Dy$_2$O$_3$ films are formed on Ge(100) substrates. The epitaxial-relationship is identified as Dy$_2$O$_3$ (110) ‖ Ge(100) and Dy$_2$O$_3$ [001] ‖ Ge[011]. Atomic force microscopy results show that the surface of the Dy$_2$O$_3$ film is uniform, flat and smooth with root mean square surface roughness of about 4.6Å. X-ray photoelectron spectroscopy including depth profiles confirms the composition of the films being close to Dy$_2$O$_3$. TEM measurements reveal a sharp, crystalline interface between the oxide and Ge.

INTRODUCTION

For the fabrication of a high speed transistor a high mobility (high-μ) Ge(100) substrate is important due to both higher electron and hole mobilities than those of Si [1 - 3]. As GeO$_2$ films are thermodynamically unstable at high temperatures, the integration of epitaxial crystalline high dielectric constant (high-κ) thin films on Ge substrates could be possible without the presence of any interfacial GeO$_2$ layers [4, 5]. High-κ dysprosium oxide (Dy$_2$O$_3$) is a promising insulator among the lanthanide metal oxides due to its low hygroscopic nature [6]. The challenge is to grow epitaxial Dy$_2$O$_3$ thin films on high-μ Ge(100) substrates because the lattice mismatch between the bulk Dy$_2$O$_3$ (10.66 Å) and the Ge substrates (5.646 Å) is about 5.6% when the double Ge unit-cell is taken into account [7]. The lattice mismatch of all rare-earth oxides (RE$_2$O$_3$) with the double Ge unit-cell is shown in Fig. 1. The linear thermal expansion coefficient mismatch between Dy$_2$O$_3$ (5.9×10^{-6}/K) and Ge (7.74×10^{-6}/K) is about - 31%.

---

[*] Corresponding author: Md. Nurul Kabir Bhuiyan; Tel.: +32 16 32 7228; Fax: +32 16 32 7983;

E-mail address: kabir.bhuiyan@fys.kuleuven.be

In this paper, epitaxial Dy_2O_3 thin films are directly grown on high-μ Ge(100) substrates by molecular beam epitaxy for the first time. Structural, chemical, surface morphological and interfacial properties of the Dy_2O_3 thin film are investigated in detail by *in situ* Reflection high energy electron diffraction (RHEED) and *ex situ* X-ray diffraction (XRD), X-ray photoelectron spectroscopy (XPS), Atomic force microscopy (AFM) and cross sectional Transmission electron microscopy (TEM).

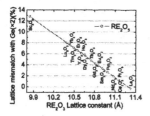

Figure 1. Lattice mismatch between rare-earth oxides and the double unit cell of Ge.

EXPERIMENT

Thin films of Dy_2O_3 are grown on Ge(100) substrates in a solid-source MBE system (DCA Instruments, Finland). This oxide-MBE system is connected to a load lock chamber for a sample transfer and is equipped with four cluster source ports (four ports per cluster) for effusion cells, electron beam guns, piezoleak valves and RHEED investigations. The base pressures for the load lock and the growth chamber are better than 2×10^{-8} and 4×10^{-10} mbar, respectively. Pure Dy metal (99.7%) is evaporated from a Knudsen cell (K-cell) and molecular oxygen (O_2) is introduced into the growth chamber through a piezoleak valve. Ga-doped p-type Ge(100) wafers with a resistivity of 0.011- 0.23 Ω-cm are used as starting substrates. The Dy growth rate – measured by a quartz crystal microbalance – is about 5.9Å/min. Ge substrates are cleaned by annealing at high temperature and Dy_2O_3 thin films are then grown on the reconstructured Ge(001)-(2×1) surfaces at a substrate temperature of about 450°C in an O_2 partial pressure of 3×10^{-7} mbar using a co-deposition method. Subsequently, the Dy_2O_3 thin films are annealed at 450°C for 20 min without O_2 in a background pressure of about 6×10^{-10} mbar for the improvement of the film crystalline quality.

The crystallinity of the film surface during the growth is monitored by *in situ* RHEED with an acceleration voltage of 15 keV (Staib Instruments GmbH, Germany). Further on, the crystallinity of the film is investigated by *ex situ* XRD using Cu-Kα X-rays (PANalytic, The Netherlands). The composition, hygroscopic and interfacial properties of the films are investigated by the depth profile XPS (Thermo Fisher Scientific) and cross sectional TEM (FEI Tecnai, 300 kV). The surface morphology of the film is observed by *ex situ* tapping-mode AFM (Digital Instruments, USA).

RESULT AND DISCUSSION

A p-type Ge(100) substrate is annealed at 550°C for 20 min in a pressure of 1×10^{-9} mbar, which leads to a smooth, chemically clean and well-ordered Ge(100)-(2×1) superstructure. The superstructure is clearly observed in the RHEED patterns taken along the Ge[110] azimuth, as shown in Fig. 2(a). Ge is a face centered cubic (fcc) crystal with a = 5.646 Å. Thus, in this orientation the most intense streaks correspond to an interplanar spacing of 3.99Å. The (2×1) reconstruction can be seen as weak streaks with a half spacing between the transmitted beam the first streak.

After deposition of Dy_2O_3 [Fig. 2(b)] on the reconstructed Ge(100)-(2×1) surface [Fig. 2(a)], the (2×1) RHEED pattern rapidly disappears and – after a critical thickness sharp and intense streaky features appear, suggesting the growth of epitaxial Dy_2O_3 films with smooth and flat surfaces (i.e., 2D films) [8, 9]. The interplanar distances of the Dy_2O_3 film corresponding to the first three steaks (their spacing is marked by red, blue and green line, respectively) are calculated to be 7.59, 3.76 and 1.87Å, respectively. The lattice constant, derived from the second, most intense streak and multiplied by 2√2, results in 10.64 Å, in agreement with the bulk Dy_2O_3 lattice parameter. Accordingly, the in-plane orientation-relationship is identified to be Dy_2O_3 [001]∥Ge[011] . The in-plane lattice mismatch between the substrate and the film at 450°C can be calculated to +5.86%. The positive sign means the lattice constant of the film is smaller than of the substrate leading to tensile epitaxial stress. Close to the most intense streaks, two other features (their spacing is marked by pink and yellow line, respectively) are found with the interplanar distances of 5.21 and 2.58 Å, respectively. These features originate from the equivalent orientation of the Dy_2O_3 unit cell on the cubic (100) Ge substrate surface. To be precise, these streaks correspond to domains which are 90° rotated with respect to the aforementioned (110) orientated Dy_2O_3 unit cell with the rotation axis parallel to the substrate surface normal. Thus, in these domains the in-plane orientation-relationship is Dy_2O_3 [001]∥Ge[0-11].

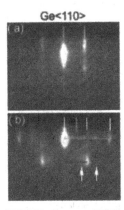
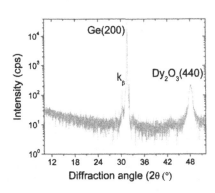

Figure 2. RHEED patterns of (a) Ge(100)-(2×1) superstructures, (b) Dy_2O_3 thin films on Ge(100)-(2×1).

Figure 3. XRD patterns of Dy_2O_3 thin films on Ge(100)-(2×1) surfaces.

Dy_2O_3 films are further investigated using XRD measurements in order to understand and identify the crystalline phases and the epitaxial-orientation relationship between the film and the substrate. Figure 3 shows the XRD pattern (typical θ - 2θ scan) of the Dy_2O_3 film grown on the Ge(100) substrate at 450°C. The Ge(200) diffraction peak from the substrate can be found at 2θ = 31.6° [8, 10]. A diffraction peak with strong intensity is observed at 2θ = 48.15° which corresponds to an out-of-plane spacing of 1.89Å in agreement with the (440) diffraction plane of the cubic Dy_2O_3. This peak is almost identical to the bulk peak position which corresponds to 2θ = 48.25° [10]. Hence, the out-of-plane epitaxial-relationship can be identified as Dy_2O_3 [110] ∥ Ge[100] in agreement with RHEED results. As the (440) Dy_2O_3 film peak is the only occurring peak, we can assume that the films are single crystalline Dy_2O_3.

The chemical composition of the dielectric Dy_2O_3 film grown on Ge(100) substrates is analyzed by XPS. An Al-Kα monochromator is used to excite the photoelectrons analysed with constant analyzer energy (CAE) mode pass energy of 100 eV. The binding energy (BE) scale is calibrated to the C 1s peak at 285 eV. XPS core level spectra of Dy $3d_{5/2}$ and O 1s lines of the Dy_2O_3 film are investigated and the peaks following the subtraction of Tougaard background are fitted, as shown in Fig. 4.

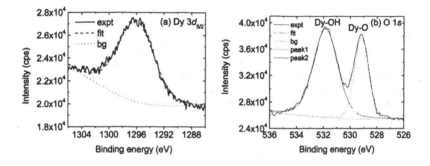

Figure 4. XPS core level spectra of (a) Dy $3d_{5/2}$ and (b) O 1s for the Dy_2O_3 film.

The peak position of the Dy $3d_{5/2}$ peak from Dy_2O_3 films is found to be at BE =1296.14 eV as shown in Fig. 4(a) [11-13]. The O 1s peak is clearly split into two clear broad peaks [Fig. 4(b)]. At lower binding energy, peak position of the O 1s peak is identified at BE=529.16 eV, corresponding to Dy-O bonding of Dy_2O_3 films [11-13]. At higher binding energy, the peak position of the O 1s peak is observed at BE=531.85 eV, corresponding to Dy-OH and Dy-C-O-bonding and possibly absorption of H_2O on the surface of Dy_2O_3 films [14, 15].

Figure 5 shows XPS sputter depth profiles of Dy_2O_3 thin films grown on Ge(100) substrates. Sputter depth profiles are carried out with 3 keV Ar^+ beams and a current density of 0.22 $\mu A/mm^2$. It is seen that the carbon contamination is present only on the surface of the film. No contaminants are observed in the Dy_2O_3 thin films [16]. The concentration of Dy and O is homogeneous inside the film, in agreement with stoichiometric Dy_2O_3. The Dy concentration exhibits a tail in the Ge substrate due to ion beam mixing effects and the lower sputtering coefficient of the heavier Dy [17]. As already stated [Fig. 5], no Ge is measured on the surface of the Dy_2O_3 film since the film is too thick (130Å). The Ge 2p peak is detectable when the Dy_2O_3/Ge interface is approached during the XPS sputter depth profile. The FWHM of the Ge $2p_{3/2}$ peak decreased from 2.1 eV when first detected (after 100s sputtering) to 1.5 eV (180s sputtering) to 1.4 eV in the bulk of the Ge substrate while the peak position is within 1217.1±0.1 eV. The eventual presence of the following Ge compounds is possible: Ge-O, Dy-Ge and Dy-Ge-O. However, the lack of reliable reference XPS data for those compounds in the literature and the effects of preferential sputtering on the stoichiometry make their identification difficult. Anyway, the relative amount of these species is low.

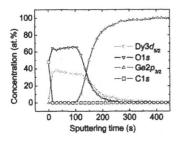 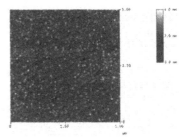

Figure 5. XPS depth profiles of the epitaxial Dy_2O_3/Ge thin film.

Figure 6. AFM image of the epitaxial Dy_2O_3/Ge thin film.

The surface roughness of the Dy_2O_3/Ge thin film using AFM is presented in Fig. 6. The scanning area and the z-height of the AFM image are 5μm × 5μm and 40 Å, respectively. The Dy_2O_3 film consists of very small grains which are related to the film crystallinity. The rms surface roughness of the Dy_2O_3 film is calculated to be 4.6Å. The surface of the film is flat and smooth.

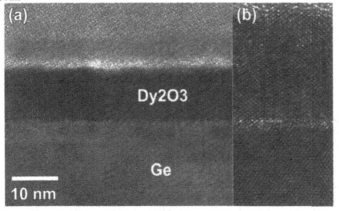

Figure 7. Cross sectional TEM micrographs of the epitaxial Dy_2O_3/Ge thin film in (a) low and (b) high magnification.

A cross sectional TEM image of the film can be seen in Fig. 7. The low-magnification image [Fig. 7(a)] indicates a crystalline film without any presence of an interfacial layer. The high resolution image [Fig. 7(b)] confirms a sharp, crystalline interface between the film and the Ge substrate. The TEM studies reveal the epitaxial relationship derived from RHEED and XRD measurements as the predominant. Columnar grains with a diameter ranging from 80 to 180Å are

31

also observed, which originate from the 90° rotation of the Dy_2O_3 unit cell around the Ge[100] axis, as has been observed previously for the growth of Dy_2O_3 on $SrTiO_3$ [18].

CONCLUSIONS

The high-κ Dy_2O_3 thin film is grown on Ge(100) substrates epitaxially by MBE and the structural, compositional, morphological and interfacial properties of the film are investigated by *in situ* RHEED and *ex situ* XRD, XPS, AFM and TEM. The Dy_2O_3 film is of the single crystalline cubic phase epitaxially grown on the high-μ Ge(100) substrate with (110)-orientation. The epitaxial-relationship is identified to Dy_2O_3 (110) ‖ Ge(100). It is seen that the surface of the Dy_2O_3 film is very flat and smooth. These structural results form an excellent basis to explore the integration of Dy_2O_3 into epitaxial electrical devices and in particular in gate stack structures.

ACKNOWLEDGMENTS

The authors gratefully acknowledge to the seventh framework program of European Project, Dual-channel CMOS for (sub)-22 nm high performance logic (DUALLOGIC), the Research Foundation Flanders (FWO) for financial support of this research activity. We thank Bastiaan Opperdoes and Dr. Stijn Vandezande for technical assistance and maintenance of the MBE system and XRD.

REFERENCES

1. A. Dimoulas, D. P. Brunco, S. Ferrari, J. W. Seo, Y. Panayiotatos, A. Sotiropoulos, T. Conard, M. Caymax, S. Spiga, M. Fanciulli, Ch. Dieker, E. K. Evangelou, S. Galata, M. Housa, M. M. Heyns, Thin Solid Films **515**, 6337 (2007).
2. T. Lee, S. J. Rhee, C. Y. Kang, F. Zhu, H. Kim, C. Choi, I. Ok, M. Zhang, S. Krishnan, G. Thareja and J. C. Lee, IEEE Electron Device Letters **27**, 640 (2006).
3. J. W. Seo, Ch. Dieker, J. -P. Locquet, G. Mavrou and A. Dimoulas, Appl. Phys. Lett. **87**, 221906 (2005).
4. Alessandro Molle, Md. Nurul Kabir Bhuiyan, Grazia Tallarida, and Marco Fanciulli, Appl. Phys. Lett. **89**, 083504 (2006).
5. Alessandro Molle, Michele Perego, Md. Nurul Kabir Bhuiyan, Claudia Wiemer, Grazia Tallarida, and Marco Fanciulli, J. Appl. Phys. **102**, 034513 (2007).
6. Sanghun Jeon and Hyunsang Hwang, J. Appl. Phys. **93**, 6393 (2003).
7. Ahn *et. al.*, United States patents, US **2005/0023594 A1** and US **2005/0124175 A1.**
8. Apurba Laha, H. J. Osten and A. Fissel, Appl. Phys. Lett. **89**, 143514 (2006).
9. Alessandro Molle, Claudia Wiemer, Md. Nurul Kabir Bhuiyan, Grazia Tallarida, and Marco Fanciulli, Appl. Phys. Lett. **90**, 193511 (2007).
10. Powder Diffraction File (PDF) database, International Centre for Diffraction Data (ICDD), File No. **PDF 04-002-5116.**
11. K. Han, Y. Zhang, T. Cheng, Z. Fang, M. Gao, Z. Xu, X. Yin, Materials Chemistry and Physics 114, 430 (2009).
12. D. D. Sarma and C. N. R. Rao, J. Electron Spectrosc. Relat. Phenom. 20, 25 (1980).
13. B. D. Paladia, W. C. Lang, P. R. Norris, L. M. Watson and P. J. Fabian, Proc. Roy, Soc. Ser. A **354**, 269 (1977).

14. C. Mahata, M. K. Bera, T. Das, S. Mallik, M. K. Hota, B. Majhi, S. Verma, P. K. Bose and C. K. Maiti, Semicond. Sci. Technol. **24,** 085006 (2009).
15. M. Perego, A. Molle and M. Fanciulli, Appl. Phys. Lett. **92,** 042106 (2008).
16. S. Toyoda, H. Kamada, T. Tanimura, H. Kumigashira, M. Oshima, T. Ohtsuka, Y. Hata and M. Niwa, Appl. Phys. Lett. **96,** 042905 (2010).
17. T. Conard, C. Huyghebaert, W. Vandervorst, Applid Surface Science **231-232,** 574 (2004).
18. A. Catana and J.-P. Locquet, Journal of Materials Research **8,** 1373 (1993).

Mater. Res. Soc. Symp. Proc. Vol. 1252 © 2010 Materials Research Society 1252-I04-06

Investigation of the Thermal Stability of Strained Ge Layers Grown at Low Temperature by Reduced-Pressure Chemical Vapour Deposition on Relaxed $Si_{0.2}Ge_{0.8}$ Buffers

A. Dobbie, M. Myronov, Xue-Chao Liu, Van H. Nguyen, E. H. C. Parker and D. R. Leadley

Nano-Silicon Group, Department of Physics, University of Warwick, Coventry, CV4 7AL, U.K.

ABSTRACT

The thermal stability of thin strained germanium (s-Ge) channels on high quality relaxed $Si_{0.2}Ge_{0.8}$ buffers has been investigated using *in-situ* H_2 annealing at temperatures between 450 °C and 650 °C. The relaxation of the s-Ge epitaxial layers was found to increase with both the s-Ge layer thickness and the annealing temperature, through a combination of surface roughening (rms surface roughness was at least 2 (7) times higher following 550 °C (650 °C) annealing compared to the as-grown s-Ge layers) and misfit dislocation formation at the s-$Ge/Si_{0.2}Ge_{0.8}$ interface. Our results suggest that the thermal budget for s-Ge device fabrication should be kept below 550 °C, in order to retain high quality s-Ge layers and minimize degradation of the carrier mobility.

INTRODUCTION

High quality strained Ge (s-Ge) layers are a promising candidate to achieve high mobility channels in MOSFETs suitable for the 22 nm technology node and beyond, due to the intrinsically higher mobility of Ge compared to Si, and the additional performance enhancements from strain [1, 2]. In order to achieve a s-Ge channel more than a few monolayers thick on a Si substrate it is necessary to engineer a relaxed $Si_{1-x}Ge_x$ buffer with a high Ge content (x > 0.5). We have recently reported high quality s-Ge layers grown by reduced-pressure chemical vapour deposition (RPCVD) at low temperature (T ≤ 450 °C) on a fully relaxed $Si_{0.2}Ge_{0.8}$ buffer [3]. By using a reverse-grading approach, we achieved a high Ge composition in the buffer, with a smooth surface (rms roughness of ~ 2 nm), low threading dislocation density (~ 4 x 10^6 cm^{-2}) and much thinner (~ 2.1 μm) than can be achieved with conventional linear grading [4].

In order to maximize the performance potential of s-Ge channel transistors, control of both the s-Ge surface passivation prior to gate stack formation and the employment of a low thermal budget to avoid relaxation of the s-Ge channel are critical [5, 6]. In this work we have investigated the thermal stability of s-Ge channels grown on relaxed $Si_{0.2}Ge_{0.8}$ buffers in H_2 ambient as a function of both the s-Ge layer thickness and the annealing temperature. Annealing temperatures in the range of 450 °C – 650 °C were used, as these temperatures are similar to those currently used during the fabrication of advanced CMOS devices.

EXPERIMENT

The s-Ge layers investigated in this study were grown by RP-CVD using an ASM Epsilon 2000 reactor on 200 mm reverse-linearly graded relaxed $Si_{0.2}Ge_{0.8}$ buffers. Figure 1 depicts the full layer structure and the growth details of the buffer are reported elsewhere [4]. The s-Ge layers (up to 80 nm thick) were grown at 400 °C using GeH_4 gaseous precursor [3]. Immediately after growth, the strained layers were subject to an *in-situ* anneal within the CVD reactor in H_2 ambient at temperatures up to 650 °C for 10 mins. With the exception of the anneal temperature, all other process conditions were kept constant, namely chamber pressure, H_2 carrier gas flow and wafer rotation speed.

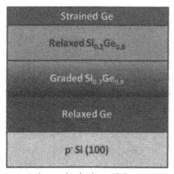

Figure 1. Schematic design of the reverse-graded relaxed $Si_{0.2}Ge_{0.8}$ buffer with a s-Ge layer.

High resolution x-ray diffraction (HR-XRD) was used to analyse the relaxation state of the annealed s-Ge layers using a Philips X'pert MRD Pro single crystal high resolution x-ray diffractometer equipped with a four-crystal Ge (220) monochromator. Reciprocal space maps (RSMs) around the symmetric (004) and asymmetric (224) reciprocal lattice points were used to determine both the *in-plane* and *out-of-plane* lattice constants of the s-Ge layer and all layers within the buffer. This allowed both the relaxation and the germanium composition of the final composition layer of the buffer to be determined independently, which is important, as this controls the amount of strain the strained layers are subjected to. The relaxation of the s-Ge layer was determined with respect to the underlying constant composition layer of the $Si_{0.2}Ge_{0.8}$ buffer.

An accurate value for the thickness of the s-Ge layer was obtained from cross-sectional transmission electron microscopy (XTEM) using a JEOL 2000FX operating at 200 kV. XTEM also enabled us to examine the formation of any defects that are generated either within the s-Ge layer or at the interface between the strained layer and the buffer following the annealing process.

The surface morphology of the annealed layers was examined by atomic force microscopy (AFM), using a Veeco Multimode AFM with a Nanoscope IIIa controller. Values of the root mean square (rms) surface roughness were determined, typically from the average of 3 scan images of at least 10×10 μm taken from different positions on the 200 mm wafer.

DISCUSSION

Figure 2 shows RSMs around the (224) reciprocal lattice point for (a) an as-grown s-Ge layer with a thickness of around 80 nm and (b) an 80 nm s-Ge layer subjected to a 10 min H_2 anneal at 650 °C. For each RSM four peaks are labelled corresponding to the Si substrate, the relaxed Ge layer (r-Ge), the relaxed $Si_{0.2}Ge_{0.8}$ layer and the s-Ge layer. The region between the relaxed Ge and $Si_{0.2}Ge_{0.8}$ peaks corresponds to the graded layer of the relaxed buffer. Each RSM shows three lines, which indicate layers that are fully strained with respect to the silicon substrate (vertical, dashed), fully strained with respect to the relaxed $Si_{0.2}Ge_{0.8}$ layer (vertical, dotted) and fully relaxed with respect to the silicon substrate (diagonal, dash-dotted), in order to demonstrate the strained state of each layer. Both the relaxed Ge and $Si_{0.2}Ge_{0.8}$ peaks lie above the dash-dotted line, indicating they are both over-relaxed with respect to Si, and hence under small tensile strain. This is due to the differences in the thermal expansion coefficients of Si and Ge [7]. The centre of the as-grown s-Ge peak lies exactly underneath the relaxed $Si_{0.2}Ge_{0.8}$ peak, confirming that the as-grown 80 nm s-Ge layer is fully strained with respect to the relaxed $Si_{0.2}Ge_{0.8}$ layer. However, after annealing at 650 °C, the s-Ge peak position has shifted towards that of the relaxed Ge peak (indicated by the arrow), meaning that the strained layer has partially relaxed. In this particular case the s-Ge layer is 62 % relaxed with respect to the relaxed $Si_{0.2}Ge_{0.8}$ layer. All relaxation values for the s-Ge take into account the effects of tilt in the s-Ge layer, which occurs during the relaxation process, and is evident on the RSMs around the (004) reciprocal lattice point (not shown) [8].

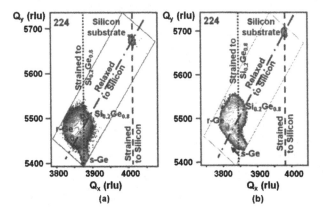

Figure 2. Reciprocal space maps around the (224) reciprocal lattice point for an 80 nm s-Ge layer on a reverse-graded relaxed $Si_{0.2}Ge_{0.8}$ buffer (a) as-grown and (b) after a 10 min *in-situ* H_2 anneal at 650 °C.

The relaxation of the s-Ge layer is plotted as a function of the s-Ge layer thickness in figure 3 for three anneal temperatures between 450 °C and 650 °C. The amount of strain remaining in the strained layer depends strongly on both the layer thickness and the annealing temperature. Strained Ge layers with a thickness of up to 50 nm remained fully strained after annealing at

450 °C, whereas after annealing at 550 °C s-Ge layers with a thickness of just 20 nm were found to be on the onset of relaxation (around 4 % relaxed with respect to the $Si_{0.2}Ge_{0.8}$ layer). Figure 3 clearly shows that an annealing temperature of 650 °C is too high, as at that temperature all of the s-Ge layers became partially relaxed, with relaxation values exceeding 20%.

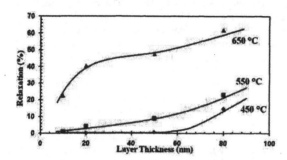

Figure 3. Relaxation of s-Ge layers as a function of Ge layer thickness after H_2 annealing between 450 °C and 650 °C. Lines are shown to guide the eye.

Figure 4 shows typical AFM images for a 20 nm s-Ge layer with increasing anneal temperature. Although the rms surface roughness only increased slightly after annealing at 450 °C, with annealing at 550 °C and 650 °C it increased to around 5 nm and 17 nm, respectively. Figure 4 also shows that as the annealing temperature is increased, there is a transition in the surface morphology from one exhibiting the typical crosshatch of the underlying relaxed SiGe buffer to one showing the formation of Ge islands.

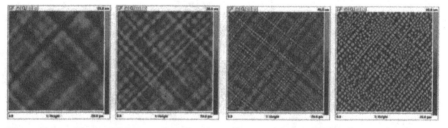

Figure 4. 20 × 20 μm AFM scans depicting the surface morphology of s-Ge layers after H_2 annealing. The annealing temperatures (from left to right) are as-grown, 450 °C, 550 °C and 650 °C, and the height ranges of the AFM scans are 20 nm, 20 nm, 40 nm and 80 nm, respectively.

The rms surface roughness is shown as a function of s-Ge layer thickness and annealing temperature in figure 5. All the annealed layers exhibited a rougher surface than the as-grown strained layers, for which the rms is ~ 2 nm [3]. Annealing the s-Ge layers at 450 °C resulted in only a slight increase in surface roughness, almost independent of s-Ge layer thickness. However, annealing at the higher temperatures above 550 °C results in significant increases in the surface roughness as the s-Ge layers undergo relaxation. Furthermore, at these higher annealing temperatures, the surface roughness exhibits a stronger dependence on the s-Ge layer thickness, such that the thickest s-Ge layers were found to be smoother.

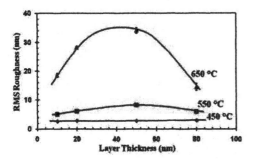

Figure 5. RMS surface roughness of strained Ge layers as a function of layer thickness after *in-situ* H_2 annealing at different temperatures. The RMS surface roughness of as-grown strained Ge layers is 2 nm (not shown) [3]. Lines are shown to guide the eye.

Figure 6 shows XTEM images of an 80 nm s-Ge layer (a) as-grown and (b) after annealing at 650 °C. The as-grown s-Ge layer has a smooth surface, a well-defined hetero-interface with the relaxed $Si_{0.2}Ge_{0.8}$ buffer layer, and no defects are observed either within the s-Ge layer or at the s-Ge/$Si_{0.2}Ge_{0.8}$ hetero-interface. Upon annealing the surface of the s-Ge layer shows visible roughening and defects are clearly seen at the s-Ge/$Si_{0.2}Ge_{0.8}$ interface. The defects at the s-Ge/$Si_{0.2}Ge_{0.8}$ interface are most likely to be misfit dislocations that are formed during the strain relaxation of the Ge layer during annealing. Such dislocations were not as readily observed in the thinner s-Ge layers, indicating that the strain relaxation occurs predominantly via surface

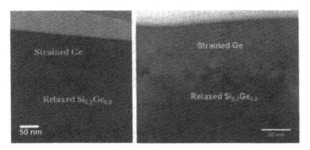

Figure 6. XTEM image of an 80 nm s-Ge layer on a relaxed $Si_{0.2}Ge_{0.8}$ buffer, showing the as-grown layer (left) and the layer following H_2 anneal at 650 °C.

roughening, whereas for the thicker s-Ge layers both surface roughening and misfit dislocations at the s-Ge/Si$_{0.2}$Ge$_{0.8}$ interface contribute to the strain relaxation. The additional contribution to the strain relaxation from misfit dislocations could help explain why the thickest s-Ge layers were found to be smoother (Figure 5).

CONCLUSIONS

We have studied the thermal stability of epitaxial s-Ge layers grown on a relaxed Si$_{0.2}$Ge$_{0.8}$ buffer by annealing in H$_2$ ambient at temperatures between 450 °C and 650 °C. The relaxation of the s-Ge layers was found to exhibit a strong dependence on both the channel thickness and the annealing temperature. Relaxation of the s-Ge layers resulted in a combination of both increased surface roughness and the formation of misfit dislocations the s-Ge/Si$_{0.2}$Ge$_{0.8}$ interface, both of which are can act as a source of increased carrier scattering and lower mobility. These results are important for the fabrication of s-Ge devices, and in particular for the conditions used for the passivation of the s-Ge surface prior to gate stack formation, which is critical to the performance of the CMOS device. Based on the results presented here, we have found that the thermal budget should be kept below 550 °C to avoid relaxation and roughening of the s-Ge channel, which could degrade both the carrier mobility and transistor performance.

ACKNOWLEDGMENTS

This work was supported by the EPSRC *"Renaissance Germanium"* project number EP/F031408/1 and the NANOSIL Network of Excellence, funded by the European Commission 7th Framework Programme (ICT-FP7, n°216171).

REFERENCES

1. M. Lee, C. W. Leitz, Z. Cheng, A. J. Pitera, T. Langdo, M. T. Currie, G. Taraschi, E. A. Fitzgerald and D. A. Antoniadis, *Appl. Phys. Lett.* **79**, 3344 (2001).
2. M. Myronov, K. Sawano, Y. Shiraki, T. Mouri and K. M. Itoh, *Appl. Phys. Lett.* **91**, 082108 (2007).
3. M. Myronov, A. Dobbie, V. A. Shah and D. R. Leadley, *EMRS* (2009).
4. V. A. Shah, A. Dobbie, M. Myronov and D. R. Leadley, *J. Appl. Phys.* **107**, 064304 (2010); V. A. Shah, A. Dobbie, M. Myronov, D. J. F. Fulgoni, L. J. Nash and D. R. Leadley, *Appl. Phys. Lett.* **93**, 192103 (2008).
5. H. Shang, M. M. Frank, E. P. Gusev, J. O. Chu, S. W. Bedell, K. W. Guarini and M. Ieong, *IBM J. Res. & Dev.* **50**, 377 (2006).
6. M. Caymax, F. Leys, J. Mitard, K. Martens, L. Yang, G. Pourtois, W. Vandervorst, M. Meuris and R. Loo, *J. Electrochem. Soc.* **156**, H979 (2009).
7. Y. Ishikawa, K. Wada, D. D. Cannon, J. Liu, H.- C. Luan and L. C. Kimerling, *Appl. Phys. Lett.* **82**, 2044 (2003).
8. M. Erdtmann and T. A. Langdo, *J. Mater. Sci: Mater.Electron.* **17**, 137 (2006).

Mater. Res. Soc. Symp. Proc. Vol. 1252 © 2010 Materials Research Society   1252-I04-07

High electron mobility in Ge nMISFETs with high quality S/D formed by solid source diffusions

Tatsuro Maeda[1], and Yukinori Morita[1] and Shinichi Takagi[2]

[1] Nanodevice Innovation Research Center - National Institute of Advanced Industrial Science and Technology (NIRC-AIST), AIST Tsukuba Central 4, Tsukuba, Ibaraki 305-8562, Japan
[2] The University of Tokyo, 7-3-1 Hongo, Bunkyo-ku, Tokyo 113-8656, Japan

ABSTRACT

We fabricate high-k/Ge n-channel MISFETs with the gate-last process to realize both of high quality source/drain (S/D) and gate stack. The n^+/p junction formed by thermal solid source diffusion of Sb dopant exhibits the excellent diode characteristics of $\sim 1.5 \times 10^5$ on/off ratio between ± 1 V and the quite low reverse current density of $\sim 4.1 \times 10^{-4}$ A/cm^2 at $+1$ V after the fabrication of high-k/Ge n-channel MISFETs that enable us to observe well-behaved transistor performances. The extracted electron mobility with the peak of 891cm^2/V·s is high enough to be superior to Si universal electron mobility especially in low E_{eff}.

INTRODUCTION

Recently Ge has regained tremendous attention because of its higher hole and electron mobility than those of Si [1]. Compared with Ge p-channel MISFETs, however, poor electrical characteristics of Ge n-channel MISFETs have been reported in spite of improved subthreshold slopes (SS), smaller surface leakage currents, and lower interface state densities (D_{it}) [2-5]. And there is not yet a clear understanding of the causes behind the degraded electron mobility. Interestingly, most of the electron mobility data reported previously were extracted from n-channel MISFETs with significantly high S/D series resistance (R_{SD}) and poor n^+/p diode characteristics. Actually, it is a great challenge to obtain high quality S/D in the gate-first process since the activation temperatures for S/D formations in Ge should usually be very low in order to avoid the degradation of Ge gate stacks and device isolations. Thus, as far as using the gate-first process, it is quite hard to characterize the precise electron mobility in the company of high quality gate stack and S/D with the clear device isolation. In order to mitigate this thermal constraint of Ge gate stacks and surface passivations, the gate-last process was implemented for Ge n-channel MISFET fabrications recently [6,7]. However, S/D junction properties were still insufficient due to unoptimized S/D fabrication process. Further improvements in n^+/p diode properties are desired for precise characterizations of Ge n-channel MISFETs. Thus, in this study, we employ the gate-last process in conjunction with the solid source impurity diffusion techniques for n^+ regions instead of the conventional ion implantation technique. Thanks to the excellent n^+/p diode characteristics such as low R_{SD} and small junction leakage, we successfully realize the decent Ge n-channel MISFET operation to extract precise intrinsic Ge electron mobility.

EXPERIMENT

Figure 1 schematically shows the fabrication process of Ge n-channel MISFETs with three mask steps. The Ge substrates used for device fabrication were In-doped p-type (100) substrates with the resistivity of 0.34Ωcm. In order to fabrication of n^+ S/D regions, the solid source diffusion techniques of Sb were employed instead of the conventional ion implantation technique because we can avoid the generation of crystal defects, redundant dopants, and dopant precipitates. It has been pointed out that most of electron mobility data reported previously are extracted from Ge n-channel MISFETs using ion implantation techniques. There is possibility that if ion implantation damages are not repaired by low-temperature annealing process, these damaged sites might provide dangling bonds and act electron traps as an accepter [8]. It seems that this p-type damage region may cause high R_{SD} to connect between S/D and channel as well as to make trouble to electron injections and emissions at S/D edge. The solid source diffusion technique has an advantage to keep away from these problems. After the diffusion windows of SiO_2 films were opened, heavily Sb-doped silicate glasses were coated. Sb dopant from Sb-doped silicate glass layer diffused at 700°C for 1 hour in a vacuum furnace. After removing SiO_2 films, HfO_2 films were deposited as a gate insulator [9]. Following the contact hole etching, metal gate electrodes overlapping the S/D region and metal contacts for S/D were formed by a lift-off process. Post metallization annealing (PMA) was carried out at the temperatures of 200-350°C for 30min in atmospheric inert gas. In the final device structure, the device isolation between S/D and channel regions was done by n^+/p junctions (Fig. 1 (h)). It must be noted that all Ge surfaces except contact hole areas were passivated by HfO_2 films and the field passivation layers and gate stacks were subjected to thermal treatment only of PMA.

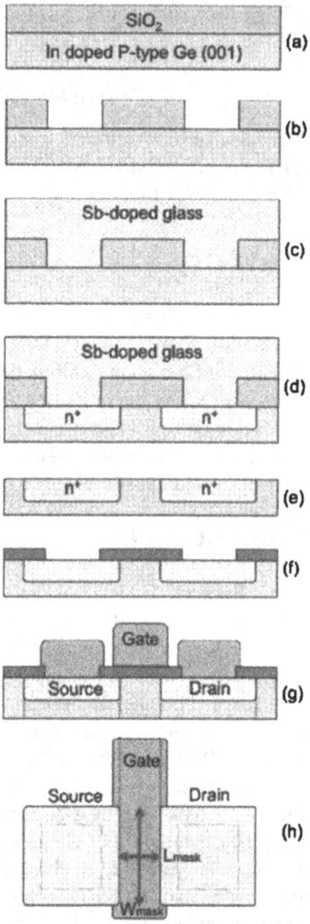

Fig. 1   Process flow of Ge n-channel MISFETs with simple three mask steps.

RESULTS AND DISCUSSION

Formation of high quality n^+ region

It is very important to investigate the basic physical properties of the fabricated n^+ region because n^+/p diode is a fundamental component in n-channel FET along with a gate stack. From the secondary ion mass spectroscopy (SIMS) analysis (Fig. 2), the depth of resultant n^+/p junction is around 2.5μm and the maximum surface concentration of Sb is 5×10^{18}cm^{-3}, which is nearly solid solubility of Sb in Ge [10]. Hall sheet carrier concentration of 8.73×10^{14}cm^{-2} for Sb diffused region is almost consistent with Sb dose from SIMS measurement, meaning the full dopant activation of Sb. The electron Hall mobility of 607cm^2/V·s exhibits a reasonable value in comparison with those in the Ge single crystal [11]. These data are excellent proofs of the

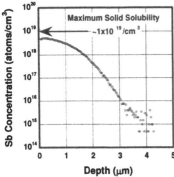

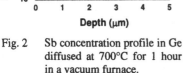

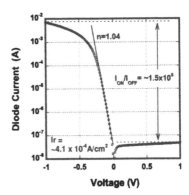

Fig. 2   Sb concentration profile in Ge diffused at 700°C for 1 hour in a vacuum furnace.

Fig. 3   I-V junction characteristic of n^+/p diode of 100μm/100μm area .

formation of high quality n^+ region. Also, the sheet resistance of 11.6 Ω/sq is low enough to observe the decent MISFET operation.

I-V characteristics of the n^+/p diode

Figure 3 shows I-V junction characteristics of the fabricated n^+/p diode of 100μm/100μm area. The excellent diode characteristics exhibiting ~1.5 x 10^5 on/off ratio are achieved between ±1 V. It is noteworthy that the reverse current density is quite low level of ~4.1 x 10^{-4}A/cm^2 at +1 V which is among the best values reported. This is possibly attributed to the reduction of the surface leakage current and bulk generation current in depletion layer due to the appropriate Sb diffusion process and the effective Ge surface passivation by HfO$_2$ films. In consequence of the low reverse current, we can observe an obvious ideal diode region in the forward current where the ideal factor of 1.04 is close to unity. These results indicate the formation of the outstanding high quality n^+/p junction. From the curve fitting of the forward diode current, the series resistance is extracted to be around 100 Ω. Given the resistivity of n^+ region and p-type substrate, the contact resistance is composed mainly of the series resistance.

MISFETs characteristics

Thanks to the ideal diode characteristics of S/D area, we have successfully realized the decent Ge n-channel MISFET operation. The typical characteristics of Ge n-channel MISFETs are displayed in Fig. 4. The channel length/width of mask (L_{mask}/W_{mask}) is 100μm/100μm defined by S/D region as shown in Fig.1 (h). Figures 4 show measured drain/substrate current - gate voltage (I_D/I_{SUB}-V_G) (Fig. 4 (a)) and source current - gate voltage (I_S-V_G) (Fig. 4 (b)) characteristics at drain voltage (V_D) = 0.01 and 1V. The ratio of I_{Don}/I_{Doff} is as low as 2.7×10^2 at V_D = 0.01V and subthreshold slope (SS) is 289mV/dec attributed to the high I_{Doff}. Here, we have found that I_{Doff} is dominated by I_{SUB} that corresponds to the reverse junction leakage in the drain area. This fact is corroborated by the finding that the values of I_{Doff} are consistent with a reverse diode current as shown in Fig. 3 because both size of n^+ region are the same. Also, it is confirmed that measured I_D is equivalent to the sum of I_S and I_{SUB}. Therefore, measured I_S can be regarded as the actual channel current. I_{Soff} is suppressed to the order of 10^{-10}A that is critical

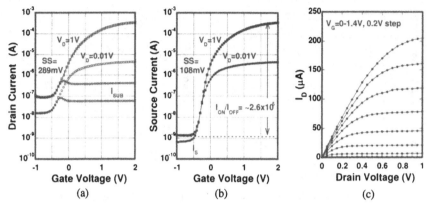

Gate Voltage (V)
(a)

Gate Voltage (V)
(b)

Drain Voltage (V)
(c)

Fig. 4 I_D/I_{SUB}-V_G (a), I_S-V_G (b) characteristics at V_D = 0.01 and 1V and I_D - V_D characteristics (c) of Ge n-channel MISFETs. L_{mask}/W_{mask} is 100μm/100μm defined by S/D region as shown in Fig. 1 (h).

factor in the accuracy of the extracted mobility. On account of low I_{Soff}, the ratio of I_{Son}/I_{Soff} amounts to 6.3×10^3 and 2.6×10^5 at V_D = 0.01 and 1V, respectively and the SS value of 108mV/dec is achieved without any drain induced barrier lowering. The interface state density (D_{it}) derived from SS is about 3.0×10^{12}cm^{-2}·eV^{-1}. It is interesting to note that I_{SUB} becomes constant in accumulation, shows in peaks in subthreshold region and becomes constant again in inversion region. This behavior is also reasonable that I_{SUB} recognizes the reverse current of gate controlled n$^+$/p junction [12]. Fig. 4 (c) shows well-behaved I_D-V_D characteristics with flat I_D in the saturation region.

Electron mobility extraction

To accurately calculate the carrier mobility, the effective channel width (W_{eff}) and length (L_{eff}) should be derived appropriately. The dependence of the peak transconductance (Gm) on W_{mask} for L_{mask}=100μm is shown in Fig. 5 (a). ΔW, which is the difference between W_{mask} and W_{eff} can be determined clearly as the intercept on x axis of liner fitting of Gm at different W_{mask}. Extracted ΔW is 20μm, which is quite large due to the large overlap along the gate width in the gate area. Figure 5 (b) shows L_{eff} and R_{SD} extracted by plotting the externally measured total device resistance (R_{tot}) vs L_{mask} for W_{mask}=100μm. From the common intercept of liner lines for a different V_G-V_{th}, the values of R_{SD} and the difference between L_{mask} and L_{eff} (ΔL) are evaluated to be approximately 400Ω and 5μm, respectively. The obtained L_{eff} shows a reasonable value for the results from the diffusion length of Sb dopant as mentioned before. R_{SD} of 400Ω is quite low compared to previous data although main component of R_{SD} is expected to be contact resistance between metal and n$^+$ region as explained before. It is apparent that the effect of R_{SD} on R_{tot} is significantly small in low V_G but is indispensable in high V_G. Hence, R_{SD} should be taken into considerations to estimate intrinsic electron mobility.

44

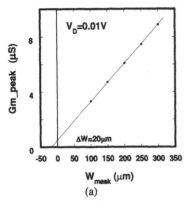
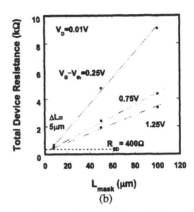

W_{mask} (μm)
(a)

L_{mask} (μm)
(b)

Fig. 5   The dependence of the peak transconductance (Gm) on W_{mask} for L_{mask}=100μm (a) and R_{tot} vs L_{mask} for W_{mask}=100μm (b)

Using appropriate IV and split CV measurements with effective parameters, electron mobility (μ_{eff}) as a function of the effective electric field (E_{eff}) for different PMA temperatures are extracted as shown in Fig. 6 (a). The devices treated with higher temperature PMA show the superior electron mobility. The impact of R_{SD} in the actual electron mobility at PMA temperature of 350 °C are shown in Fig. 6 (b). It is obvious that the effect of R_{SD} on μ_{eff} is significantly small in low E_{eff} but is crucial in high E_{eff}. Hence, R_{SD} should be taken into considerations to estimate intrinsic electron mobility. Consequently, we can observe higher Ge electron mobility than Si universal mobility at low E_{eff} and obtain the highest electron mobility with the peak mobility of 891cm^2/V·s. To our knowledge, this is one of the highest mobility reported on high-k/Ge devices. The extracted mobility without R_{SD} is still relatively high compared with that reported before. Therefore, the formation of high quality S/D rather than high R_{SD} has great impact to obtain high electron mobility. However, the extracted electron mobility is still lower than expected, possibly due to relatively high D_{it}. Thus, by realizing high interface quality of gate stacks, the electron mobility should be further improved.

CONCLUSIONS

In conclusion, we fabricate high-k/Ge n-channel MISFETs with the gate-last process in conjunction with the solid source impurity diffusion techniques for n$^+$ regions. Thanks to the formation of high quality S/D with the effective Ge surface passivation by HfO$_2$ films, we can observe higher Ge electron mobility than Si universal mobility at low E_{eff} and obtain the highest electron mobility with the peak mobility of 891cm^2/V·s. It is found that the formation of high quality S/D has great impact to obtain high electron mobility. As a result, the low electron mobility in high-k/Ge n-channel MISFETs, reported so far, is not necessarily limited by high S/D series resistance or high D_{it}, but much higher electron mobility can be expected by further improving the device fabrication processes of S/D and gate interface qualities.

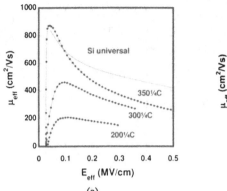
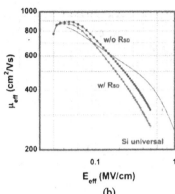

(a)   (b)

Fig. 5   Electron mobility as a function of the effective electric field (E_{eff}) for different PMA
temperatures and the difference of electron mobility with and without R_{SD} at PMA
temperature of 350°C (b).

ACKNOWLEDGMENTS
    The authors would like to thank Dr. T. Kanayama for their support and encouragement.
The device fabrication was carried out at the AIST Nanoprocessing Facility. This work was
supported by the New Energy and Industrial Technology Development Organization (NEDO).

REFERENCES
1. S. Takagi, VLSI Tech. Dig. 2003, 115 (2003).
2. S. J. Whang, S. J. Lee, F. Gao, N. Wu, C. X. Zhu, J. S. Pan, L. J. Tang, and D. L. Kwong,
    Tech. Dig. - Int. Electron Devices Meet. 307 (2004).
3. G. Nicholas, D.P. Brunco, A. Dimoulas, J. Van Steenbergen, F. Bellenger, M. Houssa, M.
    Caymax, M. Meuris, Y. Panayiotatos, A. Sotiropoulos, IEEE Trans. Electron Devices 54,
    1425 (2007).
4. Jin-Hong Park, Munehiro Tada, Duygu Kuzum, Pawan Kapur, Hyun-Yong Yu, H-.S. Philip
    Wong, Krishna C. Saraswat, Tech. Dig. - Int. Electron Devices Meet. 389 (2008).
5. J. Oh, I. Ok, C.-Y. Kang, M. Jamill, S.-H. Lee, W.-Y. Loh, J. Huang, B. Sassman, L. Smith,
    S. Parthasarathy, B. E. Coss, W.-H. Choi, H.-D. Lee, M. Cho, S. K. Banerjee, P. Majhi, P. D.
    Kirsch, H.-H. Tseng and R. Jammy, VLSI Tech. Dig. 2009, 238 (2009).
6. K. Mirii, T. Iwasaki, R. Nakane, M. Takenaka, and S. Takagi, Tech. Dig. - Int. Electron
    Devices Meet. 457 (2009).
7. C. H. Lee, T. Nishimura, N. Nagashio, K. Nagashio, K. Kita, and A.Toriumi, Tech. Dig. - Int.
    Electron Devices Meet. 681 (2009).
8. J. R. Weber, A. Janotti, P. Rinke, and C. G. Van de Walle, Appl. Phys. Lett. 91, 142101
    (2007)
9. T. Maeda, Y. Morita, M. Nishizawa, and S. Takagi, ECS Trans. 3, 551 (2006).
10. F. A. Trumbore, The Bell System Technical Journal 205 (1960).
11. C. Hilsum, Electron. Lett. 10, 259 (1974).
12. A. S. Grove, *Physics and Technology of Semiconductor Device*, p. 298, (Wiley, New York,
    1967)

Poster Session

Mater. Res. Soc. Symp. Proc. Vol. 1252 © 2010 Materials Research Society 1252-I05-05

Investigation of Wet Etch of sub-nm LaOx Capping Layers for CMOS applications

Hui-feng Li, Mo Jahanbani, Martin Rodgers, Steve Bennett, Daniel Franca, Corbet Johnson, Steven Gausepohl, Joseph Piccirillo, Jr.

College of Nanoscale Science and Engineering, SUNY AT Albany, 255 Fuller Road, Albany, NY 12203

ABSTRACT

With the continuous scale-down of Si CMOS devices to sub-45nm region, high permittivity gate dielectric and metal gate stacks have begun to replace the traditional SiO_2 gate oxide and poly Si gate. The introduction of HfO_2 gate dielectric and metal gate stack raised new challenges in modulating MOSFET threshold voltage. LaOx cap layer has been shown to be a promising candidate for work function tuning to reduce the NFET threshold voltage in low standby power CMOS devices. During the processing, HfO_2 gate dielectrics on PFET areas need to be excluded from LaOx deposition and only the HfO_2 gate dielectrics on NFET area is exposed to LaOx deposition. After LaOx diffusion into HfO_2 on NFET area, LaOx on top of PFET needs to be etched in the subsequent processing. The subsequent etch requires removal of LaOx and underneath protective mask on PFET and keep the HfO_2 gate dielectrics on NFET active area undamaged. This poses a big challenge as to the choice of the etch process. The underneath mask also had an impact on the subsequent LaOx etch. In this paper we investigated wet etch of LaOx using amorphous Si and TiN as the underneath blocking layer. TiN is a more suitable mask candidate than amorphous Si for LaOx etches. XPS showed LaOx diffused into amorphous Si and made the subsequent a-Si etch difficult.

INTRODUCTION

The pursuit of high performance and low power in Si microelectronics has resulted in continuously featuring size shrinking of CMOS technology. As the gate length is scaled down, the gate oxide thickness also needs to scale down to maximize the control of gate electrode over the channel and improve the short channel effects [1]. The scaling of gate oxide caused an increase in gate leakage. The gate leakage becomes significant as the gate length is scaled down to below 90nm node [1]. The scaling difficulty encountered in silicon oxide/oxynitride requires the replacement of the conventional SiO(N) with gate dielectrics with a high permittivity as the technology enters into sub-45nm node. With a HfO_2 dielectric material, the gate dielectric can be physically deposited thicker than conventional SiO(N) at an effectively equivalent oxide thickness (EOT). Moreover, using HfO_2 as the gate dielectrics provides an opportunity of continuously scaling down the gate dielectrics with lower EOT, leading to higher MOSFETs performance [2, 3].

Poly silicon has conventionally been used as the gate electrode. Various threshold voltages can be reached by doping poly Si electrodes to be n-type for nFETs and p-type for pFETs for high performance applications. The poly gate electrode effectiveness has been impacted significantly as the gate dielectric scaling and the introduction of HfO_2 gate dielectrics due to poly depletion and interaction of HfO_2 gate dielectrics with existing poly Si gates [4]. The interaction caused high interface trap densities that pinned the threshold of a MOSFET to an undesirable value. Fully silicided poly Si gates (FUSI) have been practiced trying to eliminate the poly depletion [5]. There are, however, concerns on FUSI scalability and non-uniform silicidatioon across different gate lengths. Metal gate has thus emerged as a choice for high performance MOSFETs in sub-45nm node [6].

To replace n+ and p+ poly Si gates and maintain low threshold voltage and good short channel effects, a band-edge metal solution is pursued. Dual metal gate electrodes with different work functions have been investigated to achieve low threshold voltages for n- and p-MOSFETS, respectively [7,8]. An alternative approach to the dual metal gate solution is to use rare-earth metal oxide capping layer on top of HfO_2 to shift the metal gate work function and thus modulate the threshold voltage of a MOSFET. Among various rare-earth metal oxides investigated, LaOx has demonstrated to be a promising candidate to modulate the threshold voltage of HfO_2/metal gate n-MOSFET transistors [9]. To integrate LaOx into the CMOS process, HfO_2 gate dielectrics in n-FET area is deposited with LaOx while p-FET area is protected by a hard mask. After La is driven into interfacial layer/HfO2 interface, LaOx on top of p-FET area is removed by wet etch. This paper will investigate the process integration of LaOx in a CMOS process.

EXPERIMENT

Two types of samples were fabricated to investigate the wet etch of LaOx. One was to deposit layer stacks on a blanket wafer and the other was patterned layer stacks for effectiveness.

After HfO_2 film was deposited by atomic layer deposition and hard mask layer was deposited on top of Hf-based HfO_2 film, a thin layer PVD LaOx was deposited on top of the hard mask. The sample was then plasma annealed in an ambient of N_2. In this investigation, TiN and a-Si were chosen as the hard mask.

The wet etch of LaOx was performed in wet tool DNS SU-3000. TXRF and XPS were employed to examine La residue after wet etch.

DISCUSSION

The CMOS integration scheme is shown in figure 1. After HfO_2 gate dielectric was deposited across the wafer, the p-FET area was masked by hard mask. Then LaOx was deposited and driven into the HfO_2/ inter layer interface by annealing in N_2 ambient. The hard mask was chosen to effectively protect HfO_2 in p-FET area from the diffusion of LaOx during the annealing. Moreover the hard mask should be removable from p-FET area. TiN and a-Si were employed as gate materials in HfO_2/metal gate MOSFETs because of their compatibility with the CMOS processes. So these TiN and a-Si were evaluated as the hard mask for LaOx deposition.

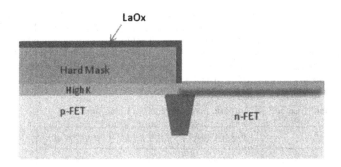

LaOx

Hard Mask

High K

p-FET

n-FET

Figure 1 Cross section of CMOS using LaOx as the capping layer

A. a-Si as Hard Mask

Total Reflection X-Ray Fluorescence (TXRF) was employed to evaluate La on the surface of the a-Si hard mask. Blanket wafers with stack of HfO_2 and a-Si layer were then deposited with LaOx. After annealing, the sample was etched with SC1 ($NH_4OH:H_2O_2:H2O$ = 3:1:1 ~ 12:1:1) for a specified time. SC1 etch was performed at the ration of (4~10):1:1 ($DI:H_2O_2:NH_4OH$) in a hot bath in DNS wet etch tool. The variation in La on top of a-Si hard mask was monitored by TXRF after each process step as shown in Figure 2.

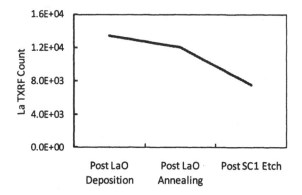

Figure 2 La TXRF count at each process step. The high La count after SC1 etch indicating significant La residue on top of a-Si hard mask

The data shows significant La residue after SC1 etch. Table 1 shows the La atomic percentage at the a-Si surface etched for extensive SC1 time. The SC1 etch was spanned to the maximum bath time in the tool. The data shows a decrease in La atomic percentage when the sample was etched initially for 3~6 min. and then became saturate even at extensive SC1 etch time for 9~12 min. The a-Si hard mask was then etched in hot NH₄OH solution to expose the underneath HfO₂. The sample was then characterized against the deposited HfO2 thickness using KLA-Tencor CD-200 UV tool. The measurement was far from the deposited HfO₂ thickness implying a-Si was not etched away completely from the HfO₂ surface in p-FET area.

Table 1 La 3d5 XPS atomic percentage vs SC1 etch time

	Pre SC1 Etch	SC1 3~6 min	SC1 9~12 min
La3d5 Atomic%	22.3	15.1	16.2

The fact that the surface La saturated at an extensive SC1 etch time implies that La might have diffused into a-Si hard mask after LaOx annealing. XPS depth profile was characterized on sample with SC1 etch of 9~12 minutes. The results are shown in Figure 3. We can see clearly that La did diffuse into a-Si hard mask and accumulated at a point very close to the a-Si surface. The La accumulation at a shallow region within a-Si film layer made the subsequent a-Si etch by NH₄OH solution very difficult. The experimental results shown above demonstrated that a-Si is not a suitable hard mask to protect p-FET area when LaOx capping layer is used to modulate the n-FET threshold voltage.

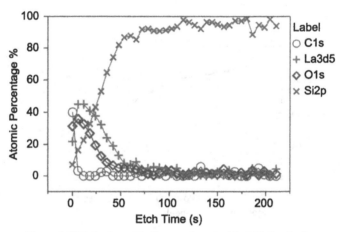

Figure 3 XPS depth profile for sample etched in SC1 for t2 min

B. TiN as Hard Mask

TiN appears as the hard mask due to the fact that TiN has been investigated as the gate material. This makes the TiN deposition compatible with the CMOS process. Another important fact is that the both TiN and LaO can be etched by SC1 solution. Thus even if LaOx diffused into the TiN layer, it could still be able to be stripped by SC1. Fig.4 shows TXRF La count when TiN was used as hard mask.

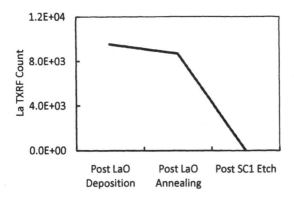

Figure 4 La TXRF count at each process step. After SC1 etch, La count is close to aero indicating complete removal of LaOx

To confirm the complete removal of LaOx and TiN hard mask after SC1 etch, the HfO2thickness was measured using KLA-Tencor's ASET-F5 thin film measurement system. The results showed the expected HfO2thickness when TiN was used as hard mask.

CONCLUSIONS

This investigation shows TiN is a suitable hard mask material on top of p-FET area when LaOx is used as capping layer to modulate the n-FET threshold voltage. Due to the fact that La diffusion into a-Si during the processing, it makes a-Si hard mask hard to be etched in the subsequent processes.

ACKNOWLEDGEMENTS

The authors would like to thank Mr. Richard Moore of the College of Nanoscale Science and Engineering of University at Albany for assistance in running XPS and data analysis.

REFERENCES

1. A. Buchanan, Scaling the gate dielectric: Materials, integration, and reliability. IBM J. Res. Dev, 43, 245 (1999)
2. G. D. Wilk, R. M. Wallace, and J. M. Anthony, High-κ gate dielectrics: Current status and materials properties considerations, J. Appl. Phys. 89, 5243 (2001)
3. Howard R. Huff and David C. Gilmer, High dielectric constant materials: VLSI MOSFET applications
4. E.P. Gusev, V. Narayanan, M.M. Frank, Advanced high-k dielectric stacks with poly-Si and metal gates: Recent progress and current challenges, IBM J. Res. Dev, 50, 387 (2006)
5. W. Maszara, Z. Krivokapic, P. King, J. Goo, and M. Lin, Transistors with dual work function metal gates by single full silicidation (FUSI) of polysilicon, IEDM Tech. Dig., pp. 367–70
6. R. Chau, S. Datta, M. Doczy, B. Doyle, J. Kavalieros, and M. Metz, High-k/Metal–Gate Stack and Its MOSFET Characteristics, IEEE Electron Device Lett. 25, 408 (2004)
7. S. C. Song, Z. B. Zhang, M. M. Husain, C. Huffman, J. Barnett, S. B. Bae, H. J. Li, P. Majhi, C. S. Park, and B. S. Ju, Highly Manufacturable 45nm LSTP CMOSFETs Using Novel Dual High-k and Dual Metal Gate CMOS Integration, VLSI Symp. Tech. Dig., 13 (2006)
8. C. Ren, H. Yu, H. J. F. Kang, X. P. Wang, H. H. H. Ma, Y.-C. Yeo, D. S. H. Chan, M.F. Li, and D. –L. Kwong, A Dual-Metal Gate Integration Process for CMOS With Sub-1-nm EOT HfO_2 by Using HfN Replacement Gate, IEEE Electron Device Letters, 25, 580 (2004)
9. C. Y. Kang, P. D. Kirsch, B. H. Lee, B. H Tseng, and R. Jammy, Reliability of La-Doped Hf-Based Dielectric nMOSFETs, IEEE Trans. Device Mater. Rel. 9, 171 (2009)

Mater. Res. Soc. Symp. Proc. Vol. 1252 © 2010 Materials Research Society   1252-I05-08

Profiling different kinds of generated defects at elevated temperature in both SiO$_2$ and high-k dielectrics

S. Sahhaf^{a,b}, R. Degraeve^a, M.B. Zahid^a, and G. Groeseneken^{a,b}

^{a} IMEC, Kapeldreef 75, B-3001, Heverlee, Belgium
^{b} KULeuven, ESAT Department, Leuven, Belgium

Corresponding author: Sahar Sahhaf, IMEC, Kapeldreef 75, B-3001 Heverlee, Belgium. Tel.: +32 (0) 16 28 76 69.
E-mail: sahar.sahhaf@imec.be

ABSTRACT

In this work, the effect of elevated temperature on the generated defects with constant voltage stress (CVS) in SiO$_2$ and SiO$_2$/HfSiO stacks is investigated. Applying Trap Spectroscopy by Charge Injection and Sensing (TSCIS) to 6.5 nm SiO$_2$ layers, different kinds of generated traps are profiled at low and high temperature. Also the Stress-Induced Leakage Current (SILC) spectrum of high-k dielectric stack is different at elevated temperature indicating that degradation and breakdown at high temperature is not equivalent to that at low temperature and therefore, extrapolation of data from high to low T or vice versa is challenging.

INTRODUCTION

Due to internal heating, the operating temperature of advanced CMOS technology is considerably higher than ambient room temperature. Consequently, several stress-induced degradation phenomena that limit the circuit reliability are accelerated as compared to room temperature evaluations on test structures.

Understanding the temperature dependence of oxide degradation under uniform stress conditions is generally accepted as a critical issue that should be accounted for in reliability projections. Especially in gate stacks that contain high-k materials, the number of published studies on the temperature dependence of degradation is still limited.

In this work, we characterize traps generated at elevated temperature in both SiO$_2$ and a selected SiO$_2$/HfSiO gate dielectric stack. The obtained results will make us alert about the limitations in extrapolating the data (e.g. breakdown time) from test temperature to another temperature.

We demonstrate that the generation of oxide defects changes as a function of temperature. Degradation and breakdown at high temperature is therefore not equivalent to that at low temperature. Similar observations have already been reported by Kaczer et al. on single layer SiO$_2$ gate dielectric [1], using indirect argumentation and which is confirmed by the trap characterization results presented here. As a consequence, straightforward extrapolation of Stress-Induced Leakage Current (SILC) and Time-Dependent Dielectric Breakdown (TDDB) data from high to low T or vice versa is problematic.

EXPERIMENTAL

The first group of samples characterized in this work consists of electrically-stressed NMOS transistors fabricated at IMEC using a CMOS process with conventional 6.5 nm gate oxide films. The second group of samples is fabricated using a standard high-k/MG integration flow [2]. 3nm HfSiO (50% Hf) was deposited by Atomic layer deposition (ALD) on a chemical SiO_2 interface layer resulting from IMEC clean [3]. DPN / PNA were done to incorporate 9% N in the film. TaN was used as metal gate. As a result, a 1nm SiO_2/3nm HfSiO(N) was formed with EOT of 1.71nm. The transistors show state of the art performance.

Effect of elevated temperatures on the generation of defects in SiO_2

In this section we investigate the effect of temperature on the generation of defects under constant voltage stress (CVS) in a single layer SiO_2. In order to clearly observe the consequences of the temperature increase, we characterize the energy spectrum of the defects generated during stress at the two extreme temperatures in our measurement window i.e. at 50°C and 200°C.

One of the common techniques for defect profiling is SILC spectroscopy [4] where the defect spectrum is obtained by plotting the ratio of the change in leakage current (after stress) to the initial tunnel current as a relative metric for defect generation. The peaks in the SILC spectrum are then attributed to different energetically distributed defects. Notably, in thick SiO_2 layers, SILC spectroscopy technique cannot reveal the true energy profile of defects because of the following reasons:

1) At sense voltages (low), the initial tunnel current cannot be accurately measured (within the noise level). Also extrapolating the current measured at higher voltages (following Fowler-Nordheim mechanism) to lower voltages is incorrect.

2) At low voltages used for sensing the defects, the energetically shallow traps (created at high stress voltages) are not accessible resulting in incomplete defect scanning.

In order to obtain detailed information on the energy profile of defects in 6.5 nm SiO_2, we use Trap Spectroscopy by Charge Injection and Sensing (TSCIS) [5]. The principle of TSCIS is illustrated in **Fig.1** for a 2-layer stack and explained as follows: a charging voltage (V_{charge}> V_{th}) is applied to the gate of an nmos transistor as a function of charging time (t_{charge}). During this time, traps inside the bulk of the gate dielectric are charged by direct tunneling of electrons from the inversion layer. In between the charging intervals, the gate voltage is switched for short times (~3ms) to a sense voltage (V_{sense}< V_{th}) and source-to-drain current is measured. The drop of I_{SD} compared to an initially measured I_{SD}-V_G is then converted into a V_{th}-shift. The measured voltage shift for increasing t_{charge} and V_{charge} is then transformed into a trap density map vs. energy and spatial position, using a self-consistent combination of the WKB-approximation for tunnelling and a Poisson solver. All TSCIS measurements in this paper are performed at room temperature.

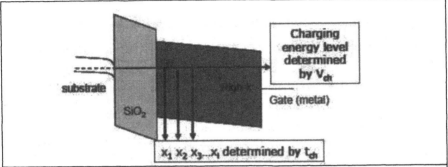

Fig.1: Illustration of the measurement principle of Trap Spectroscopy by Charge Injection and Sensing (TSCIS).

Fig.2 shows the defect profiles of the stressed SiO_2 devices at 50°C and 200°C obtained by TSCIS. We observe that when stress is performed at low temperatures, more energetically deep traps are generated compared to energetically shallow traps. This is not the case for devices stressed at high temperatures wherein homogeneously distributed defects in a wide energy range are identified. This was indirectly also observed by Kaczer *et al.* in SiO_2 [1] as they report that oxide defects created at different temperatures are not equivalent. We remark that the differences between the defects generated at elevated temperatures are even more pronounced in the high-k stack as will be discussed in next section. Consequently, degradation and breakdown at high temperature is not equivalent to that at low temperature and therefore, straightforward extrapolation of Stress-Induced Leakage Current (SILC) and Time-Dependent Dielectric Breakdown (TDDB) data from high to low T or vice versa is problematic in materials used for End-of-Roadmap Scaling of CMOS Devices.

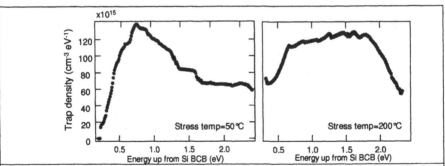

Fig.2: Defect energy profile of the NMOS devices stressed at 50°C and 200°C characterized by TSCIS. Both samples are stressed close to breakdown (BD). At low temperatures, less shallow traps are generated by CVS compared to deep traps while at high temperatures, the magnitude of the generated defect density at different energy levels is similar.

Effect of elevated temperatures on the generation of defects in High-k

In order to investigate the effect of temperature on thin $SiO_2/HfSiO$ samples, we study the SILC energy spectrum [4] of the defects generated after stressing at 75°C and 200°C. As shown in **Figs.3a** and **b**, SILC spectroscopy is obtained by plotting the gate voltage dependence of the ratio of the change in leakage current to the initial tunnel current as a relative metric for defect generation. The spectrum of a device stressed and sensed at 200°C (**Fig.3b**) shows two peaks with almost equal amplitude but different energy positions. The first peak P1 at V_g=~0.75V corresponds to energetically deep traps, while the second peak P2 at V_g=~1.3V is caused by energetically shallower traps. The spectrum at 200°C (**Fig.3b**) significantly differs from the spectrum of a device stressed and sensed at 75°C (**Fig.3a**), where only one dominant peak P1 at ~0.75V is observed.

We remark that at 75°C there might be a small amount of shallow traps that are hidden in the shoulder of the dominant SILC peak. These results are reproducible on several devices as evidenced by the data in **Fig.3c**. At 200°C, the ratio of the magnitudes of the SILC peaks P1 to P2 is close to one, while at 75°C, the ratio is ~0.4. Even when the sample that was stressed at 75°C, is afterwards heated to 200°C, the spectrum (**Fig.3d**) remains different from the one stressed at 200°C (**Fig.3b**). This proves that the difference between spectra (**3a**) and (**3b**) is not solely due to a difference in conductivity through the two trap species and consequently, that more energetically shallow traps were generated at 200°C than at 75°C. This is consistent with our results on SiO_2 discussed in previous section and also the observations of Kaczer *et al.* in SiO_2 [1] that oxide defects created at different temperatures are not equivalent and affected by the creation temperature.

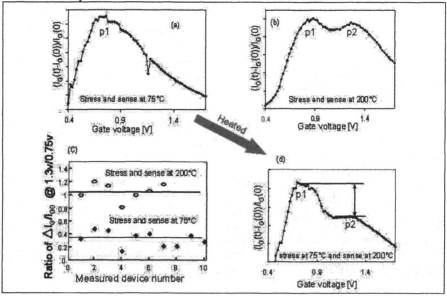

58

Fig.3: SILC spectrum of a device stressed and sensed at 200°C (b) shows two peaks P1 and P2 with almost equal amplitude but different energy, while the spectrum of a device stressed and sensed at 75°C (a) shows only one dominant peak. By repeating the experiment of several devices (c), in this stack, we see that the ratio between the trap density measured at V_g=1.3 and 0.75 is systematically close to one at 200°, while at 75°C, the ratio is close to 0.4. A spectrum of the sample that was stressed at 75°C and then heated to 200°C afterwards is shown in (d). There is still an obvious difference compared to (b), i.e., stressing at different temperatures is not equivalent.

The position of the peaks across the sense voltage range in **Fig.3** can be converted into a trap energy position in the band diagram of the stack, provided we know the spatial position of the traps. Since SILC is a trap-assisted tunneling process, the most optimal place for the traps to conduct, will be close to the SiO$_2$/HfSiO interface. At V_g=0.75V, a trap at the SiO$_2$/HfSiO interface will be aligned with the injected electrons from the substrate at ~2.75eV below the SiO$_2$ conduction band level, or, equivalently, at 1.25 eV below the HfSiO CB level. For the second SILC spectrum peak P2, we find ~2.5eV and 1eV resp.

The observed traps are not necessarily only present at the interface. Our further investigation [6] (not discussed in this work) show that the defect bands stretch out into the high-k layer as schematically shown in **Fig.4.**

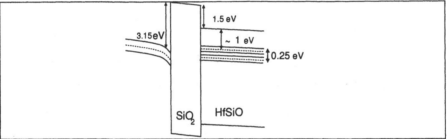

Fig.4: Band diagram for 1nm SiO$_2$/ 3nm HfSiO gate stack. The traps causing P1and P2 are localized at the interface between the SiO$_2$ and HK and stretch out into the HK.

CONCLUSIONS

In this work we investigated the effect of temperature on the generation of defects under constant voltage stress (CVS) in a single layer SiO$_2$ and SiO$_2$/HfSiO stack. Applying TSCIS on SiO$_2$, at low temperature, more energetically deep traps (neutral defects which can be charged negatively) were observed compared to energetically shallow traps. This was not the case for devices stressed at high temperature where homogeneously distributed defects in a wide energy range were identified.

A trap spectroscopy based on SILC identified more pronounced differences between the defects generated at low and high temperatures in a High-k dielectric stack. We explained that the peaks in the SILC spectrum are attributed to different energetically distributed defects. At low temperature, energetically deep traps are dominant and cause the degradation of the stack while at high temperature, more energetically shallow traps are generated. The shallow traps also participate in the leakage current and the breakdown process.

Based on the observed differences between the generated traps at elevated temperatures in both SiO_2 and HK, we conclude that the degradation and breakdown at high temperature is not equivalent to that at low temperature and therefore, straightforward extrapolation of Stress-Induced Leakage Current (SILC) and Time-Dependent Dielectric Breakdown (TDDB) data from high to low T or vice versa is problematic.

ACKNOWLEDGMENTS

This work is part of IMEC's Industrial Affiliation Program, funded by IMEC's core partners: Intel, Texas Instruments, Micron, NXP, ST, Matsushita, TSMC, Samsung and Elpida. Support from IMEC's p-line for processing and AMSIMEC for electrical characterization is greatly acknowledged.

REFERENCES

1. B. Kaczer, R. Degraeve, N. Pangon, G. Groeseneken, "The Influence of Elevated Temperature on Degradation and Lifetime Prediction of Thin Silicon-Dioxide Films", IEEE Trans. On Electron Dev., vol.47, no.7, pp.1514-1521, 2000.
2. T. Schram, L. Ragnarsson, G. Lujan, W. Deweerd, J. Chen, W. Tsai, K. Henson, R. Lander, J. Hooker, J. Vertommen, C. De Meyer, S. De Gendt, M. Heyns, "Performance improvement of self-aligned HfO2/TaN and SiON/TaN nMOS transistors", Microelectronics reliability, vol.45, pp.779-782, 2005.
3. M. Meuris, P.W. Mertens, A. Opdebeeck, H.F. Schmidt, M. Depas, G. Vereecke, M.M. Heyns and A. Philipossian, "The IMEC clean: A new concept for particle and metal removal on si surfaces", Solid State Technology, vol.387, pp. 109-113, 1995.
4. R. O'Connor, L. Pantisano, R. Degraeve, T. Kauerauf, B. Kaczer, Ph. J. Roussel, G. Groeseneken, "SILC defect generation spectroscopy in HfSiON using constant voltage stress and substrate hot electron injection", Proceedings IRPS, pp.324-329, 2008.
5. R. Degraeve, M. Cho, B. Govoreanu, B. Kaczer, M.B. Zahid, J. Van Houdt, M. Jurcak and G. Groeseneken, "Trap Spectroscopy by Charge Injection and Sensing (TSCIS): a quantitative electrical technique for studying defects in dielectric stacks", Technical Digest International Electron Devices Meeting - IEDM, pp. 775-778, 2008.
6. S. Sahhaf, R. Degraeve, M. Cho, K. De Brabanter, Ph.J. Roussel, M.B. Zahid, G. Groeseneken," Detailed Analysis of Charge Pumping and I_dV_g Hysteresis for Profiling Traps in SiO_2/HfSiO(N)", submitted to Microelectronic engineering.

Mater. Res. Soc. Symp. Proc. Vol. 1252 © 2010 Materials Research Society   1252-I05-09

Electrical characteristics of crystalline Gd$_2$O$_3$ film on Si (111): impacts of growth temperature and post deposition annealing

G. Niu [1], B. Vilquin [1], N. Baboux [2], G. Saint-Girons [1], C. Plossu [2], G. Hollinger [1]

[1] INL-UMR5270/ECL, Ecole Centrale de Lyon, 36 Avenue Guy de Collongue, Ecully Cedex, 69134, France

[2] INL, INSA de Lyon, Lyon Cedex, 69432, France.

ABSTRACT

This work reports on the epitaxial growth of crystalline high-k oxide Gd$_2$O$_3$ on Si (111) by Molecular Beam Epitaxy (MBE) for CMOS gate application. Epitaxial Gd$_2$O$_3$ films of different thicknesses have been deposited on Si (111) between 650°C~750°C. Electrical characterizations reveal that the sample grown at the optimal temperature (700°C) presents an equivalent oxide thickness (EOT) of 0.73nm with a leakage current density of 3.6×10^{-2} A/cm^2 at $|V_g$-$V_{FB}|$=1V. Different Post deposition Annealing (PDA) treatments have been performed for the samples grown under optimal condition. The Gd$_2$O$_3$ films exhibit good stability and the PDA processes can effectively reduce the defect density in the oxide layer, which results in higher performances of the Gd$_2$O$_3$/Si (111) capacitor.

INTRODUCTION

The continuous scaling of the gate dielectric thickness requires high-k metal oxide as an alternative to SiO$_2$ for future CMOS (Complementary Metal Oxide Semiconductor) technology. Recently, the epitaxy of crystalline oxides on silicon attracted intensive researches: the epitaxial nature of these oxides allows circumventing the recrystallization issue encountered with amorphous oxides, which leads to a tremendous augmentation of the leakage current. In addition, using epitaxial growth techniques allows subtly monitoring the oxide stoichiometry.

Gadolinium oxide (Gd$_2$O$_3$), which belongs to the family of rare-earth (RE) metal oxides (lanthanide oxides) possesses a cubic bixbyite Mn$_2$O$_3$ (II) structure in which the unit cell includes eight unit cells of an incomplete fluorite structure. It is identified as one of the most promising candidates due to its[1] i) high dielectric constant of 20, ii) high bandgap of 5.3eV, iii) thermodynamical stability on silicon even at high temperature, and iv) very small lattice mismatch(only -0.46%) with Si (one Gd$_2$O$_3$ unit cell on two Si unit cells). Gottlob et al. reported for the first time a fully functional n-MOSFET with a TiN/Gd$_2$O$_3$/Si (001) system [2]. However, Gd$_2$O$_3$ exhibits bidomain structure on Si (001), which could significantly increase the leakage current [3]. At the same time, due to the highly perfect crystallinity of Gd$_2$O$_3$ on Si(111) and its potential application as template to integrate semiconductor (Si and Ge) [4,5] or ferroelectric/multiferroic materials with hexagonal structure (such as YMnO$_3$) [6] , Gd$_2$O$_3$/Si(111) system attracts wide research interests. Several good electrical results of this system have been reported [7,8]. Nevertheless, the reliability and stability of the Gd$_2$O$_3$ dielectric still remains acute and requires further investigations.

In the present work, we demonstrate a good quality epitaxial single domain growth of Gd_2O_3 films on p-type Si (111) substrates, with abrupt interfaces. Structural characterizations of the Gd_2O_3 films were performed using X-ray Diffraction (XRD) and X-ray Reflectivity (XRR). The dielectric properties of C-V and I-V were extracted from C-V and I-V measurements on an Au/Ni/Gd_2O_3/Si MOS structure. The impact of the growth temperature is studied. For the sample deposited at optimal temperature, Post Deposition Annealing (PDA) in different conditions has been performed to improve its quality and to investigate the stability of the system.

EXPERIMENT

All the Gd_2O_3 films were eptaxially grown on p-type (boron doped, 1×10^{16} B/cm^3) silicon substrates in a RIBER 2300 MBE reactor modified for the growth of oxides. A Quartz Crystal Microbalance (QCM) is employed to monitor in-situ the growth rate. The samples were fabricated at different growth temperatures (650°C~750°C) and the oxygen supply was finely tuned to prevent the formation of silica or silicate layers at the interfaces. More growth details can be found in Ref. [9]. At the optimal temperature of 700°C, samples with different thicknesses were prepared. The crystallographic quality and thicknesses of the samples were determined *ex-situ* by XRD and XRR respectively, using a Rigaku Smartlab diffractometer. Post Deposition Annealing (PDA) in different conditions, including a 200°C PDA in a tubular furnace and a 400°C Rapid Thermal Annealing (RTA) under oxygen or nitrogen atmosphere were performed for the samples grown at optimal temperature to improve their electrical quality and investigate their structural stability. A lift-off technique was employed to fabricate MOS capacitors, by depositing Au (250 nm)/ Ni (3 nm) metal electrodes with an area of $100 \times 100\mu m^2$. The C-V and I-V curves were respectively measured by an HP4284A and an HP4156B at room temperature. Electrical parameters such as EOT and flatband voltage (V_{FB}) were extracted by fitting experimental curves with TCV simulations [10,11], taking into account quantum effects in the silicon substrate.

DISCUSSION

Structural analysis by XRD and XRR

Figure 1 shows a 2θ-ω scan around the Si 111 Bragg reflection for a Gd_2O_3/Si (111) sample grown at 700°C. The intense and sharp peak observed at 2θ=28.44° corresponds to the Si 111 substrate reflection. According to the fitting curve, the Gd_2O_3 222 peak is located at 2θ=28.87° with a Full Width at Half Maximum (FWHM, Δ(2θ)) ~1.2°. The Gd_2O_3 222 peak appears at the right side of Si (111), indicating that Gd_2O_3 is under tensile strain and that the growth remains pseudomorphic. Pendellösung fringes appear at both shoulder of the Gd_2O_3 peak, indicating a high uniformity, flatness and crystal quality of the Gd_2O_3 film.

The thicknesses of all samples were measured using XRR. Figure 2 shows the experiment and fitting results for one of these samples. The good agreement between fitting and experiment allows extracting a film thickness of 6.73nm with a surface roughness of ~0.5nm.

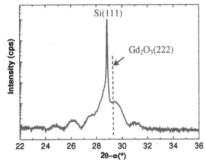

Figure 1. X-ray 2θ-ω scan around the Si (111) reflection of a 6.7nm-thick Gd_2O_3 layer grown on Si (111).

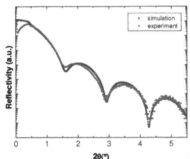

Figure 2. X-ray Reflectivity results (red dots) of 6.7nm-thick Gd_2O_3 layer together with the simulation curve (black dots).

Electrical properties

Figure 3 shows the C-V (room temperature, at 100kHz) and I-V characteristics of Gd_2O_3 layers grown at 650°C~720°C (for more details, see Ref. [9]) and 700°C turns out to be the optimal temperature since the sample deposited at 700°C shows the smallest EOT 0.73nm with a leakage current of $3.6×10^{-2}A/cm^2$ at $|V_g-V_{FB}|=1V$, which is in good agreement with the recommendation of International Technology Roadmap for Semiconductors (ITRS) for the 32nm node[12]. Figure 4 exhibits the EOT values as a function of physical thickness for the films deposited at 700°C. From the slope of the plot, the dielectric constant of the Gd_2O_3 films grown on p-type Si(111) was k=9~12, which is much less than that of bulk Gd_2O_3 (20) [1] but is very close to that of Gd_2O_3 thin films grown on Si(001) substrate (k=12~14).[13,14] Different types of charges existing in the metal/insulator/Si (MIS) system might play an important role to the variation of its dielectric properties.

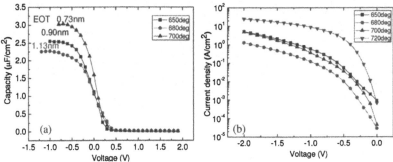

Figure 3. (a) Room temperature C-V measurements at f=100kHz for the Au/Ni/Gd$_2$O$_3$/Si(111) structures grown at different temperatures. (b) J-V measurements.

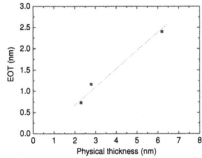

Figure 4. Evaluation of the EOTs of the Gd$_2$O$_3$ films as a function of the physical thickness.

The total charge stored in a MIS capacitance is the sum of the fix interface charge Q_f, the charge trapped on interface states Q_{it}, the mobile charge Q_m and the charge trapped in the oxide film Q_{ot}. To deduce the impact of these charges to the dielectric properties of Gd$_2$O$_3$ films grown at optimal temperature, different PDA processes have been performed. Figure 5 (a) shows the forward and reverse bias sweeps (at 100 kHz) for a 2.1nm-thick Gd$_2$O$_3$ film, both as-deposited sample and that annealed at 200°C under O$_2$ atmosphere for 30min in a tubular furnace. A clockwise loop (ΔV=0.1V) can be observed for the measurement of as-deposited sample, indicating a chargement phenomenon depending on the applied voltage [15] which are probably caused by oxygen vacancies. The flat band voltage (V$_{FB}$) extracted from the C-V curve of as-deposited sample is 0.5V, which is much larger than the theoretical V$_{FB}$ (-0.5V) of the Ni gate MOS capacitors without any oxide charge. The positive shift (ΔV$_{FB}$=1V) between the theoretical and experimental V$_{FB}$ values reveals a large amount of negative charges in the dielectric films and at the interface. The measurement of the annealed sample displays a significant decrease of the hysteresis (ΔV=0V) and the ΔV$_{FB}$ (0.7V), indicating that this PDA method under O$_2$ atmosphere effectively decreased the charge density of the as-deposited layer and developed the interface states at the Gd$_2$O$_3$/Si interface. It is very likely that the oxygen vacancies are refilled during the annealing process. At the same time, the EOT is only slightly disturbed, given that the EOT of the PDA sample is 0.9nm which is a little larger than that of as-deposited one (0.73nm). Figure 5 (b) shows the J-V measurements. At $|V_g-V_{FB}|=1V$, the leakage current density of as-deposited and PDA samples are $2.0\times10^{-2}A/cm^2$ and $4.3\times10^{-1}A/cm^2$ respectively, both of which are consistent with the recommendation of ITRS for 32nm node.

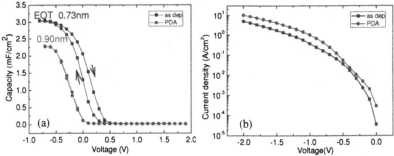

Figure 5. (a) C-V measurements at f=100 kHz for the as-deposited Gd_2O_3 film grown at 700°C and the same sample treated by a 200°C PDA process under O_2 in a tubular furnace. (b)corresponding J-V measurements.

A 400°C Rapid Thermal Annealing (RTA) under nitrogen atmosphere has also been performed for Gd_2O_3 films grown at 700°C. Figure 6 (a) shows the C-V curves measured at 100 kHz for as-deposited and RTA samples. Neither of the sweeps displays hysteresis loop indicating no chargement process occurs. The V_{FB} extracted for as-deposited and RTA samples are -0.2V and 0V respectively. The RTA sample demonstrates a significant decrease of the EOT value (3.1nm) compared to as-deposited one (4.2nm), which indicates that the RTA under N_2 atmosphere effectively reduced the defects density in the Gd_2O_3 layer resulting in an improvement of the dielectric quality. The J-V measurement shown in Figure 6 (b) confirms this better performance : at $|V_g-V_{FB}|=1V$, the leakage current density of RTA sample (2.5×10^{-2} A/cm^2) is less than that of as-deposited sample ($4.2\times10^{-2}A/cm^2$).

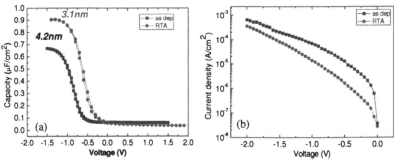

Figure 6. (a) C-V measurements at f=100 kHz for the as-deposited Gd_2O_3 film grown at 700°C and the same sample treated by a 400°C RTA process under N_2. (b) corresponding J-V measurements.

CONCLUSIONS

In conclusion, the epitaxial Gd_2O_3 films were grown on p-type Si (111) substrate by MBE. The thicknesses of the samples are determined by different methods. The sample deposited under optimal growth condition shows a dielectric constant k=9~12 and its good dielectric property is in consistent with the recommendation of the ITRS for 32nm technology. The Gd_2O_3 films demonstrate good stability when undergone different PDA treatment, 200°C under O_2 in a tubular furnace or 400°C RTA under N_2. Furthermore, these PDA treatments lead to the decrease of the defects density thus a higher performance of the Gd_2O_3/Si (111) MIS system. Gd_2O_3 has been proved to be one of the most promising candidates for crystalline gate dielectric materials of next CMOS technique generation.

ACKNOWLEDGMENTS

This work realized on the NANOLYON platform was partly supported by the French Agence Nationale pour la Recherche (ANR). The authors would like to thank S. Pelloquin, J. Penualas, P. Regreny for for helpful discussions and C. Botella, J. B. Goure for technical assistance.

REFERENCES

1. D.P. Norton, Mat. Sci. Eng. R **43**, 139 (2004)
2. H.D.B. Gottlob et al., Solid-State Electron. **50** 979-985 (2006)
3. A.Laha, E. Bugiel, J. X. Wang, Q. Q. Sun, A. Fissel, and H. -J. Osten, Appl. Phys. Lett. **93**, 182907 (2008)
4. A. Laha, E. Bugiel, A. Fissel and H -J. Osten, Microelectron. Eng. **85**, 2350 (2008)
5. G. Niu, L. Largeau, G. Saint-Girons, B. Vilquin, J. Cheng, O. Mauguin, G. Hollinger, "Monolithic integration of germanium on Gd_2O_3/Si (111) compliant substrate by molecular beam epitaxy", submitted to J. Appl. Phys. (2009)
6. Jan-Willem Bos, Bas B. van Aken, Thomas T.M. Palstra, Chem. Mater., **13**, 4804-4807 (2001)
7. J. Kwo, M. Hong, A. R. Kortan, K. L. Queeney, Y. J. Chabal, R. L. Opila, Jr., D. A. Muller, S. N. G Chu, B. J. Sapjeta, T. S. Lay, J. P. Mannaerts, T. Boone, H. W. Krautter, J. J. Krajewski, A. M. Sergnt, and J. M. Rosamilia, J. Appl. Phys. **89**, 3920, (2001)
8. A. Laha, H.-J. Osten, and A. Fissel, Appl. Phys. Lett. **89**, 143514 (2006)
9. G Niu, B. Vilquin, N. Baboux, C. Plossu, L. Becerra, G Saint-Girons, G Hollinger, Microelectron. Eng. **86**, 1700 (2009)
10. C. Busseret, N. Baboux, C. Plossu, and A. Poncet, Proceedings of SISPAD, (unpublished), 188. (2006)
11. P. Palestri et al., IEEE Trans. Electron Devices **54**, 106 (2007)
12. Http://www.itrs.net/
13. H. D. B. Gottlob et al. IEEE Electon Device Lett. **27**, 814 (2006)
14. Y. Li, N. Chen, J. Zhou, S. Song, L. Liu, Z. Yin and C. Cai, J. Cryst. Growth, **265**, 548 (2004)
15. S. M. Sze, Physics of Semiconductor Devices, Wiley, New York, (1981)

Mater. Res. Soc. Symp. Proc. Vol. 1252 © 2010 Materials Research Society   1252-I05-12

Nanoscale study of the influence of atomic oxygen on the electrical properties of LaAlO₃ thin high-k oxide films deposited by molecular beam epitaxy

Wael Hourani, Liviu Militaru, Brice Gautier, David Albertini, Armel Descamps-Mandine, Sylvain Pelloquin, Carole Plossu and Guillaume Saint-Girons

Lyon Institute of Nanotechnology, University of Lyon, UMR CNRS 5270, 7 Avenue Jean Capelle F-69621 VILLEURBANNE Cédex France

ABSTRACT

In this paper, thin films of LaAlO₃ deposited by Molecular Beam Epitaxy (MBE) using two different protocols have been studied at the nanoscale by Tunneling Atomic Force Microscope (TUNA). The aim of this study is to describe at the nanoscale the influence of the ambient gas (molecular or atomic oxygen) on the electrical properties of LaAlO₃ proposed as a gate oxide replacing SiO₂ in Metal-Oxide-Semiconductor Structures. 3 nm thick LAO films were deposited on p-type Si (100) substrates (10^{15} cm^{-3}) in an oxide-dedicated MBE reactor by electron beam evaporation of crystalline LAO targets at a substrate temperature of 400 °C in a controlled molecular O₂ or atomic O ambient (O₂ or O pressure ranging from 10^{-8} to 10^{-5} Torr). Current maps evidence the beneficial effect of the atomic oxygen ambient on the electrical properties of the film: films prepared with atomic oxygen show a lower density of "hot spots" i.e. areas of the surface where the leakage current is high for a given applied voltage. Moreover, the threshold voltage (the voltage for which the leakage current is higher than 1 pA) is higher when atomic oxygen is used. The nanoscale observations are in accordance with macroscopic intensity-voltage measurements which also show lower leakage currents when atomic oxygen is used.

INTRODUCTION

The miniaturization of the metal oxide semiconductor MOS devices following the Moore's law has lead to the extreme thinning of the commonly used SiO₂ gate oxide. However, this thinning has reached its limits because of the dramatic increase of the leakage current through the oxide, causing the degradation of the devices. Therefore, alternative high dielectric constant (high-k) oxides have been studied to replace the classical SiO₂ oxide.

Amorphous LaAlO₃ (LAO) high-k oxide appears as an interesting candidate in replacing the SiO₂ oxide; it presents a high band gap of 5.6 eV, a large conduction band offset of 1.8 eV with respect to Si and a dielectric constant of 25 in its bulk crystalline phase. Moreover, LAO is stable in air and is theoretically thermally stable in contact to silicon up to 1000 °C [1].

On the other hand, degradation and breakdown under electrical stress is one of the important reliability concerns of gate oxides. Since the breakdown phenomenon of oxides is a highly localized phenomenon [2,3,4], the microscopic characterization may provide additional information compared to the macroscopic characterization. Therefore the atomic force microscope (AFM) appears as a natural tool to characterize high-k oxides since its probe tip area is in the same order of magnitude as the breakdown spot, allowing to avoid the problem of short circuits which can take place between macroscopic electrodes and the Si substrate during the characterization of oxides having many leakage spots (the case where the measured leakage current arises from a few leakage spots cannot be distinguished from a uniform leakage current flowing from the whole electrode area).

Leakage current maps using AFM can be obtained by two different modes: Tunneling AFM (TUNA) and Conductive AFM (C-AFM). TUNA mode, which is used in our work is a very sensitive mode by which we can measure currents ranging from 100 fA to 120 pA, while by the C-AFM mode, currents between 10 pA and 100 nA are measured (this corresponds to the denomination of the manufacturer of our apparatus – Veeco instruments). The principle of the TUNA mode is the same as C-AFM, by which simultaneous topographic imaging and current imaging (constant bias is applied between the tip and the sample) or spectroscopy (Intensity-Voltage curves at a local fixed position) can be collected. TUNA operates in contact mode by using a conductive tip made of silicon coated by PtIr$_5$. Just like contact mode AFM, the z-feedback loop uses the DC cantilever deflection as a feedback signal to maintain a constant force between the tip and the sample to generate the topography image. In almost all the studies on silicon, the tip is grounded and the substrate is negatively biased to minimize possible damages of the tip by ionic radicals and to avoid anodic oxidation of the surfaces under study [5]. The principle of the TUNA is shown in the Figure 1.

TUNA technique is a powerful tool to correlate the electrical properties with the physical characteristics of the structures. Moreover, one can determine the characteristics of degradation and breakdown of ultra-thin oxides with repetitive Ramped Voltage Stress (RVS) where the applied voltage is ramped while the current flowing through the sample is recorded, leading to intensity-voltage (I-V) curves.

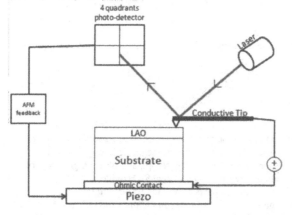

Figure 1. The principle of the TUNA mode AFM.

EXPERIMENT

Thin LAO films were deposited on p-type Si (100) substrates (10^{15} cm^{-3}) in an oxide-dedicated molecular beam epitaxy (MBE) reactor by electron beam evaporation of crystalline

LAO targets at a substrate temperature of 400 °C in a controlled molecular O_2 or atomic O ambient (O_2 or O pressure ranging from 10^{-8} to 10^{-5} Torr). In this study, two different samples have been studied: for sample A, LAO was deposited within molecular oxygen ambient, while for sample B it was deposited within atomic oxygen ambient. The physical thickness of the samples was measured using the X-ray Reflectometry (XRR): both samples have a thickness of 3 nm with incertitude of 0.3 nm.

Topographic images of the samples have been recorded using the AFM in tapping mode where the tip scans the surface area while oscillating (tapping) with a resonant frequency ranging from 280 kHz to 320 kHz. The tips used for the tapping mode AFM were made of Si. Although the topography can be obtained in TUNA mode, the AFM tapping mode allows us to obtain a higher spatial resolution (the tips are not coated with metal, so that their apex is smaller).

Current maps have been recorded simultaneously with topography in contact mode using the TUNA mode of the AFM. Spectroscopic data (Current-Voltage) have been obtained by stopping the tip over a precise region of the sample and applying repetitive RVS. Note that all the current and voltage values in the spectroscopic curves are in absolute values (the measured values provided by the system is negative due to the application of a negative voltage on the sample).

DISCUSSION

Topographies of the samples using the tapping mode AFM are shown in Figure 2.

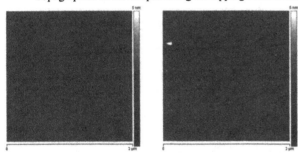

Figure 2. Topographies of the two different LAO samples, sample A (left) and sample B (right). Note that the same lateral scale has been used for the two images (2×2 μm^2).

The roughness of the surface of both samples expressed by its root mean square value (RMS) is 0.10 nm and 0.23 nm for the samples A and B respectively. So with these small values of RMS we can consider that the current measurements are not disturbed by the topography of the surface.

Current images of samples A and B are shown in Figure 3. We have applied −5.8 V on sample A and −7.2 V on B due to different average threshold voltages for both samples (voltage above which current exceeds 1 pA), which will be discussed further. On current images, darker areas correspond to places where the leakage current is higher. We notice that the density of leaky spots (black rounded areas) which are the most conductive regions where the current through the oxide exceeds 120 pA, is smaller in sample B than in sample A even with the application of a lower voltage on A. Considering several different AFM images, the density of leaky spots can be estimated: it is about 20 x 10^8 per cm^2 for sample A and 12 x 10^8 per cm^2 for that of sample B. Considering the fact that the thickness of both samples is the same (3 nm), we attribute this difference to the use of atomic oxygen for sample B and molecular oxygen for sample A.

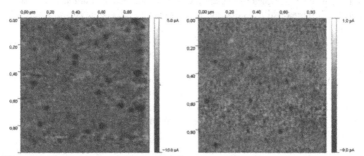

Figure 3. Current images of sample A for an applied voltage of -5.8 V (left) and of sample B for an applied voltage of -7.2V (right).

Oxygen vacancies are a serious possible problem in high-k oxides. They cause transient trapping and threshold voltage shifts. The presence of these vacancies has been studied by optical, luminescence and charge pumping experiments [6]. On the other hand, leaky spots may arise from a local thinning of the layer or the pre-existence of defects, like oxygen vacancies, allowing the current to cross the layer more easily [7]. The decrease of the density of leaky spots in the samples having LAO deposited within atomic O ambient might be related to the filling of the oxygen vacancies by the oxygen atoms, hence decreasing their density in the LAO oxide. The current image of sample B in Figure 3 confirms this influence, by the decrease of the density of the leaky spots.

We notice in the topography images (Figure 2) the existence of holes and humps. Although the hypothesis is still under study, this may be explained by this way: when it is heated during the deposition of LAO, a reaction takes place on the Si substrate, between carbon atoms which exist in the MBE chamber and the Si substrate forming silicon carbide molecules (SiC) which appear sometimes as humps and sometimes as holes when these humps explode. According to our studies, these humps and holes do not correspond to leaky spots found in current images.

To confirm the beneficial influence of the atomic O on the electrical behavior of the oxide, we have determined the I-V characteristics on 16 different regions of each sample in order to obtain the average value of the threshold voltage V_{th}. These characteristics are represented in Figure 4; the RVS are applied from 0 to -10 V with a speed of 0.5 V/s and a current limit of -80 pA to prevent the complete breakdown of the oxide (which means that the ramp is stopped when the current exceeds -80 pA).

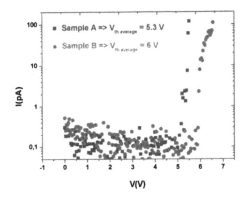

Figure 4. Representative nanoscale I-V characteristics of sample A (black curve) and of sample B (red curve) and their corresponding average values of the threshold voltages (voltage and current are represented in their absolute values).

The average value of V_{th} is 5.3 V for sample A and 6 V for sample B. This illustrates again the positive influence of the deposition of LAO oxide within atomic O ambient. The combination of this result with the lower density of leaky spots indicates that the atomic oxygen ambient should be used during the deposition of dielectric layers to enhance their electrical properties as gate oxides.

Previous results can be compared to macroscopic Intensity-Voltage (I-V) characteristics. Figure 5 shows the normalized macroscopic I-V characteristics with respect to the electrodes' surfaces on the samples A and B. The measurements have been done on electrodes with different surfaces ranging from 100 x 100 μm^2 to 600 x 600 μm^2. The good influence of the deposition of LAO within atomic oxygen ambient is also obvious from the clear difference in the density of current through samples A and B (the current density at -2 V is more than one order of magnitude for sample A with respect to sample B). From the comparison with atomic scale characterization, this can now be related to the smaller amount of nanometric leaky spots combined with a higher local threshold voltage for samples elaborated within atomic oxygen ambient.

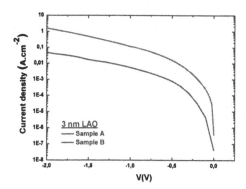

Figure 5. The macroscopic I-V characteristics of sample A (red curve) and of sample B (green curve) on 100 x 100 μm^2 electrodes.

CONCLUSIONS

In summary, thin LAO high-k oxide films were electrically characterized by TUNA mode AFM. Weak spots were found in all the samples, which correspond to local thinning of the gate oxide and to pre-existing defects in it, which may occur in the high-k oxides deposited by MBE. We have shown that the density of leaky spots is smaller and threshold voltages are higher for oxides deposited within atomic O ambient, which plays a role in reducing the density of O vacancies in the LAO oxide. In combination with macroscopic measurements, this confirms the contribution of deposition within atomic oxygen ambient to the enhancement of the properties of high-k oxides for an application as gate oxides in MOS structures.

REFERENCES

1. S. Pelloquin, L. Becerra, G. Saint-Girons, C. Plossu, N. Baboux, D. Albertini, G. Grenet, and G. Hollinger, Microelectronic Engineering, 86 (2009) 1686-1688
2. Y.-L Wu, S.-T Lin, IEEE Electronics letters, 42 (2006)
3. M. Porti, M. Nafria, X. Aymerich, A. Olbrich, and B. Ebersberger, American Institute of Physics, 78 (2001) 4181-4183
4. L. Zhang, Y. Mitani, and H. Satake, IEEE Transactions on Device and Materials Reliability, 6 (2006) 277-282
5. W. Polspoel, P. Favia, J. Mody, H. Bender, and W. Vandervorst, Journal of Applied Physics, 106 (2009) 024101-1-7
6. K. Tse, D. Liu, K. Xiong, and J. Robertson, Microelectronic Engineering, 84 (2007) 2028-2031
7. L. Yeh, I.-K Chang, C.-H Chen, and J.-M Lee, American Institute of Physics, 95 (2009) 162902-1-3

Mater. Res. Soc. Symp. Proc. Vol. 1252 © 2010 Materials Research Society   1252-I05-13

Impact of Ge Doping on Si Substrate and Diode Characteristics

J. Vanhellemont[1], J. Lauwaert[1], J. Chen[2,3], H. Vrielinck[1], J.M. Raff[4], H. Ohyama[5], E. Simoen[6] and D. Yang[2]

[1]Department of Solid State Sciences, Ghent University, B-9000 Ghent, Belgium.
[2]State Key Lab of Silicon Materials, Zhejiang University, 310027 Hangzhou, P.R. China.
[3]Institut für Angewandte Physik, TU Dresden, D-01062 Dresden, Germany.
[4]CNM-CSIC, Campus U.A.B, 08193, Bellaterra, Barcelona, Spain.
[5]Kumamoto National College of Technology, Kumamoto, 861-1102 Japan.
[6]Imec, B-3001 Leuven, Belgium.

ABSTRACT

The beneficial effects of Ge doping of Czochralski-grown Si crystals on substrate and diode characteristics are discussed and illustrated. Ge doping leads to increased oxygen precipitation during device processing, resulting in improved internal gettering in low oxygen content material while at the same time suppressing thermal donor formation. It also reduces the void size and thus the crystal originated particle size in as-grown wafers, because the Ge atoms acting as trap for vacancies close to the Si melt temperature. Furthermore Ge doping leads to an increased mechanical strength leading a.o. to reduced wafer breakage, not only during crystal wafering but also during device processing. Ge doping ($< 10^{20}$ cm^{-3}) has only a limited effect on diode characteristics and on radiation hardness.

INTRODUCTION

It is well known that with increasing crystal diameter, the interstitial oxygen concentration in Czochralski-grown Si crystals is decreasing, while in vacancy-rich crystals the size of grown-in voids - and thus also of the Crystal Originated Particles (COP's) observed on the wafer surface- is increasing, accompanied by a COP density decrease.

The first effect is related to the large melt in which movements have to be controlled and partly suppressed by the use of magnetic fields. The reduced melt movement leads to more uniform dopant incorporation but at the same time to a more limited transport of oxygen from the quartz crucible to the melt and the crystal. The lower interstitial oxygen concentration and thermal budget of modern device processing lead to a strongly reduced oxygen precipitation and thus also strongly reduced internal gettering capacity.

The increasing COP size accompanied by a decreasing density is due to the decreasing pulling rate and thermal gradient in order to avoid dislocation formation. Due to the lower thermal gradient, the vacancy concentration is decreased leading to a slower build-up of the vacancy supersaturation during cooling and thus to a lower void nucleation rate while at the same time the thermal budget for void growth controlled by vacancy diffusion is increased.

The standard approach to suppress void formation is by hot zone design to obtain the critical ratio of the pulling rate over the thermal gradient, leading to an intrinsic point defect lean crystal so that no vacancy or self-interstitial clustering occurs during crystal cooling, see e.g. [1] and

73

references therein. A drawback of this approach is that the process window is rather narrow and imposes the use of magnetic fields for Si crystals with diameter larger than 200 mm.

Another, more recent approach to reduce not only the COP problem but also that of the reduced internal gettering capacity, is introducing a dopant in the Si crystal that is not electrically active but enhances oxygen precipitation and reduces the vacancy concentration available for void nucleation and growth. Nitrogen fulfils to a large extent these requirements and nitrogen doped Cz wafers are commercially available and used already on a relatively large scale.

Also Ge is a promising dopant that can be alloyed with Si over the whole composition range. Dislocation free Cz pulling of $Si_{1-x}Ge_x$ crystals is however limited to x values well below 10%. Ge doping has been reported to enhance interstitial oxygen precipitation and out-diffusion [2], suppress thermal donor (TDD) formation [3] and to influence COP density and size [4].

Recently, Londos et al. [5] performed an extensive study of the influence of Ge doping on the behavior of O and C impurity related complexes in electron irradiated Si. The observations were explained by assuming that for Ge concentrations below 10^{20} m^{-3}, Ge atoms act as temporary traps for vacancies and as such reduce the recombination rate of intrinsic point defects and Frenkel pairs. Above 10^{20} cm^{-3} an opposite behavior was observed which was assumed to be due to the formation of Ge clusters acting as recombination centers for intrinsic point defects. Vacancy trapping by Ge atoms was also claimed by Chen et al. [6] based on density functional theory calculations. Recent quenching experiments also give evidence of vacancy trapping by Ge atoms at temperatures above 1100 °C [7].

Doping with Ge has a beneficial effect on the yield strength of the Si material and will reduce dislocation nucleation. Although the strengthening effect is more pronounced for higher Ge concentrations than used in the present study, even for lower Ge concentration it was statistically shown that wafer breakage was reduced compared to standard Si wafers [8].

In the present paper the effects of Ge doping are discussed with respect to thermal donor and COP formation as well as to diode characteristics and radiation induced defects [9-11].

CZOCHRALSKI GROWTH OF GE DOPED SI CRYSTALS AND DIODE PROCESSING

Two 4" diameter, n-type CZ crystals were pulled by QL electronics. One of the crystals (GCZ) was doped with a Ge concentration of 10^{19} cm^{-3}, whereas the second crystal (CZ) was a standard one. Both crystals were grown under the same nominal pulling conditions and had a similar resistivity and interstitial oxygen content (ASTM F 121-79) as listed in Table I.

TABLE I. Main specifications of the CZ Si and GCZ Si substrates.		
Substrate	CZ Si	GCZ Si
Type	n	n
Orientation	< 100 >	< 100 >
Thickness [μm]	525 ± 15	525 ± 15
Resistivity [Ωcm]	23.7 ± 2.4	19.4 ± 1.0
Average C_{OI} [10^{18} cm^{-3}]	1.54 ± 0.08	1.34 ± 0.08
Ge concentration [cm^{-3}]	0	10^{19}
Relative TDD generation rate R [10^{-41} $cm^6 \cdot h^{-1}$]	4.75 ± 0.1	1.56 ± 0.1

P-on-n diodes, with active area of 0.25 cm^2, were processed on polished <100> oriented wafers prepared from both crystals and were subjected to thermal treatments in N_2 /H_2 or N_2 ambient at temperatures ranging between 250 and 450 °C and annealing times between 0.5 and 5h in order to study the formation of thermal donors in finished devices. Wafers from both types received also the same heat treatment to study thermal donor formation in as-grown material. The impact of Si substrate Ge doping on diode characteristics and on thermal donor formation was analyzed by means of capacitance-voltage (C-V), current-voltage (I-V) and recombination lifetime measurements based on microwave photoconductance decay (µW-PCD) [9].

IMPACT OF GE DOPING ON CRYSTAL AND WAFER DEFECTS

The Flow Pattern Defect (FPD) and Secco Etch Pit Defect (SEPD) density in the CZ Si and GCZ Si materials were investigated by etching two wafers of each type vertically inserted in Secco etchant for different times. The observed FPD and SEP densities after 30 min Secco etching are shown in Fig. 1 together with results obtained on 5" wafers taken from different positions in crystals with different Ge concentrations [10].

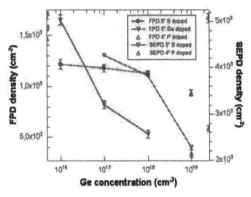
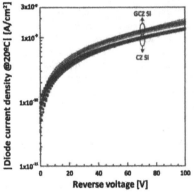

Figure 1. FPD and SEPD densities in as-grown CZ and GCZ Si crystals vs. Ge concentration. The values on the y-axes are without Ge doping [10].

Figure 2. J_{leak} vs. reverse bias for 10 CZ Si and 10 GCZ Si diodes [9].

While for the CZ Si crystals, the FPD density in the tail and head part of the crystal are quite similar, for the GCZ Si crystals, the FPD densities in the tail part are lower. This can be understood by the fact that the Ge concentration increases towards the tail of the crystal due to the segregation coefficient (0.33) of Ge in Si. Also in the 4" wafers studied in the present work, the FPD (and SEPD) density is lower in GCZ Si than in the standard CZ Si. In the present study, both for CZ Si and GCZ Si, the FPD density decreases with increasing crystal diameter while the SEPD density increases. This is the opposite behavior of the one reported before for Si [12].

The COP density was measured on 5" polished wafers which revealed an increase of COP density with increasing Ge content accompanied by a decrease of COP size [4]. On the 4" wafers a similar trend is observed but the size of the COP's in the GCZ material becomes so small that most of them are below the detection limit of the surface inspection tool.

IMPACT OF GE DOPING ON DIODE CHARACTERISTICS AND TDD FORMATION

Diodes are fabricated using a well established and stable process with high device yield [13]. Only small differences are observed between the diodes processed on the two types of substrates [9]. The reverse currents in the GCZ Si diodes are slightly higher than in their CZ Si counterparts although a somewhat higher free carrier concentration is also extracted for the GCZ Si diodes (2.20×10^{14} cm^{-3}) compared to the CZ Si ones (2.05×10^{14} cm^{-3}) (Fig. 3). This higher free carrier density is also in line with the lower resistivity as listed in Table 1. The higher reverse currents in the GCZ Si diodes are in agreement with the measured lower generation lifetimes.

An increase of the free carrier concentration is observed when subjecting the CZ Si and GCZ Si diodes to thermal anneals at 450°C (Fig. 3). After a 5 h anneal and within the probed substrate depth that corresponds to a maximum applied reverse voltage of 100 V, the carrier concentration increases by a factor of four and a factor of two for the CZ Si and GCZ Si substrates, respectively. Such increase in the carrier concentration can be attributed to the generation of oxygen-related thermal donors (TDD's). The results in Fig. 3 show that the TDD generation rate for GCZ Si ($\approx 2.6 \times 10^{13}$ cm^{-3} ·h^{-1}) is nearly 5 times lower than the one extracted for CZ Si ($\approx 1.26 \times 10^{14}$ cm^{-3} ·h^{-1}), the latter being in good agreement with previous results obtained on CZ Si substrates with similar oxygen contents. As expected there is also a slower TDD formation in a N$_2$ ambient compared to a N$_2$/H$_2$ ambient. This is explained by a H mediated increase of interstitial oxygen diffusivity resulting in a higher TDD formation rate [14].

The TDD generation rate can thus be extracted as function of the depletion depth as is illustrated in Fig. 4. Assuming that the TDD generation rate is proportional to the third power of the interstitial oxygen concentration C_{OI} one can also obtain an estimate of the C_{OI} depth profile as is shown in the same figure. The observed difference in generation rate between CZ Si and GCZ Si can be attributed to on the one hand the lower initial interstitial oxygen content in the GCZ substrates and on the other hand to the different oxygen concentration in the near surface part of the wafer due to the oxygen out-diffusion during the high temperature steps of the diode process. The TDD generation rate depends indeed on the third power of the interstitial oxygen concentration [15] while the oxygen out-diffusion is also enhanced by Ge doping in the GCZ substrates. In Fig. 4 the thus extracted TDD generation rate $r_{TDD}(x)$ depth profile is shown together with a calculated interstitial oxygen concentration depth profile after a 160 min anneal at 1100°C. The theoretical C_{OI} out-diffusion profile is approximately given by [16]

$$\frac{C_{OI}(x) - C_{OI}(0)}{C_{OI}(\infty) - C_{OI}(0)} = erf\left[\frac{x}{2\sqrt{Dt}}\right]. \tag{1}$$

x is the depth from the surface, $D = 0.13 \times \exp(-2.53eV/kT)$ the oxygen diffusion coefficient at anneal temperature T, k the Boltzmann constant and t the anneal time.

Assuming that the TDD generation rate r_{TDD} is proportional to the third power of the interstitial oxygen concentration C_{OI}, or

$$r_{TDD} = R \times (C_{OI})^3 \text{, one can write } \frac{[r_{TDD}(x)]^{1/3} - [r_{TDD}(0)]^{1/3}}{[r_{TDD}(\infty)]^{1/3} - [r_{TDD}(0)]^{1/3}} = erf\left[\frac{x}{2\sqrt{Dt}}\right], \tag{2}$$

76

and one obtains the best fits shown by the curves through the measurement points in Fig. 4. From the fit it can be concluded that the relative TDD generation rate $R = r_{TDD} \times (C_{OI})^{-3}$ is constant between 10 and 17.5 μm as listed in Table I. The agreement between the measured and calculated TDD depth profiles is excellent, illustrating that the two high temperature steps of the diode processing, i.e. the sacrificial oxidation step for 10 min at 1100°C and the field oxidation step for 145 min also at 1100°C, both in wet oxygen, are mainly determining the interstitial oxygen depth profile after the full diode process.

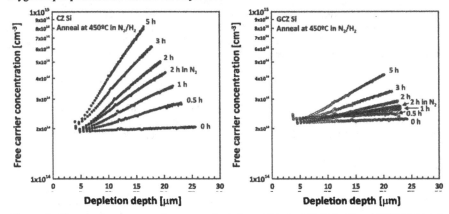

Figure 3. Free carrier concentration versus depletion width for CZ Si (left) and GCZ Si (right) diodes subjected to different thermal annealing conditions [9].

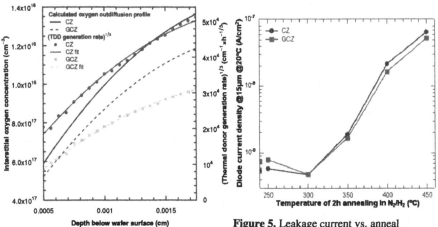

Figure 4. The TDD generation rate depth profile after anneal at 450°C, compared with the calculated C_{OI} depth profile.

Figure 5. Leakage current vs. anneal temperature for 15 μm depletion depth. Points on the y-axis are for unannealed diodes [10].

The TDD formation has also an important impact on the diode leakage current J_{leak} [9] as illustrated in Fig. 5 showing J_{leak} for a depletion depth of 15 μm as function of the 2h anneal temperature. The data points on the y-axis are for the diodes without additional anneal. A small beneficial effect of the Ge doping is observed for annealing temperatures above 300°C. It is also interesting to note that the diode leakage current improves by low temperature anneals and seems to reach a minimal value around 300°C.

Diodes were also irradiated at room temperature with 2 MeV electron fluences ranging from 10^{14} to 10^{17} e·cm^{-2} using the electron accelerator at Takasaki Japan Atomic Energy Agency. Before and after irradiation, the diode characteristics were measured with applied voltages ranging from -20 to 1 V [11]. The forward current decreases in both types of diodes after 10^{17} e·cm^{-2} irradiation, due to an increase of the resistivity of the substrate. Only a limited impact of the Ge doping on the diode characteristics is observed as is illustrated in Fig. 5.

ACKNOWLEDGEMENTS

J. Chen and J. Vanhellemont acknowledge NSFC (Grants No. 50832006 and 60906001) and FWO for financial support. Part of this work was supported by the Inter-University Laboratory for the Joint Use of JAERI Facilities. Q. Vanhellemont is acknowledged for his contribution to the simulation work.

REFERENCES

1. J. Vanhellemont, P. Spiewak, K. Sueoka and I. Romandic, Phys. Stat. Sol. C 6, 1906 (2009).
2. J. Chen, D. Yang, X. Ma, R. Fan and D. Que, J. Appl. Phys. 102, 066102 (2007).
3. C. Cui, D. Yang, X. Ma, M. Li and D. Que, Mat. Sci. Semicond. Proc. 9, 110 (2006).
4. J. Chen, D. Yang, H. Li, X. Ma, D. Tian, L. Li and D. Que, J. Cryst. Growth 306, 262 (2007).
5. C.A. Londos, A. Andrianakis, V.V. Emtsev, G.A. Oganesyan and H. Ohyama, Physica B 404, 4693 (2009).
6. J. Chen, T. Wu, X. Ma, L. Wang and D. Yang, J. Appl. Phys. 103, 123519 (2008).
7. J. Vanhellemont, M. Suezawa, I. Yonenaga, submitted for publication in J. Appl. Phys.
8. J. Chen, D. Yang, X. Ma, Z. Zeng, D. Tian, L. Li, D. Que and L. Gong, J. Appl. Phys. 103, 123521 (2008).
9. J.M. Rafí, J. Vanhellemont, E. Simoen, J. Chen, M. Zabala, F. Campabadal, Physica B 404, 4723 (2009).
10. J. Vanhellemont, J. Chen, W. Xu, D. Yang, J.M. Rafí, H. Ohyama and E. Simoen, presented at SEMICON China 2010, Shanghai March 2010, ECS Transactions, in press.
11. H. Ohyama, J.M. Rafí, K. Takakura, E. Simoen, J. Chen and J. Vanhellemont, Physica B 404, 4671 (2009).
12. T. Abe, Material Science and Engineering B 73, 16 (2000).
13. C. Martínez, J.M. Rafí, M. Lozano, F. Campabadal, J. Santander, L. Fonseca, M. Ullán and A. Moreno, IEEE Trans. Nucl. Sci. 49, 1377 (2002).
14. E. Simoen, Y.L. Huang, Y. Ma, J. Lauwaert, P. Clauws, J.M. Rafí, A. Ulyashin, C. Claeys, J. Electrochem. Soc. 156, H434 (2009).
15. P. Wagner and J. Hage, Appl. Phys. A 49, 123 (1989).
16. P. Gaworzewski and G. Ritter, Phys. Stat. Sol. (a) 67, 511 (1981).

Mater. Res. Soc. Symp. Proc. Vol. 1252 © 2010 Materials Research Society   1252-I05-13

Contact Technology Using Pulsed Laser Annealing to Form Ti/Al Ohmic Contacts on n-type GaN With Lower Contact Resistance and Improved Surface Morphology

Grace Huiqi Wang[1], Xincai Wang[2], Debbie Hwee Leng Seng[1], Hongyu Zheng[2] ,Yong Lim Foo[1] and Sudhiranjan Tripathy [1].

1. Institute of Materials Research and Engineering (IMRE), Singapore.
2. Singapore Institute of Manufacturing Technology (SIMTech), Singapore.

ABSTRACT

We employ laser annealing for metal contact formation on n type gallium nitride (n-GaN) substrate. Laser annealing helps to achieve a reduced sheet resistance in the contact formed, resulting in low resistance ohmic contact produced for high performance GaN light emitting diodes (LEDs) applications. The laser irradiation of the n-GaN layers led to an increased electron concentration and a lower sheet resistance. It was found that laser irradiation increases the electron concentration to ~5.53×10^{22} cm^{-3} as compared to ~3.18×10^{19} cm^{-3} when rapid thermal anneal (RTA) was employed. The advantage of using laser annealing, was that it reduced the barrier height more effectively than RTA, thus resulting in lower contact resistivity.0.52J cm^{-2} 5 pulses laser annealing reduced the contact resistivity of the film to 2.4×10^{-7} ohm-cm^{2} as compared to RTA at 880°C,30s[4.75 ×10^{-5} ohm-cm^{2}]. The formation of a low resistance is attributed to the increase in carrier concentration and the removal of native oxide from n-GaN by laser irradiation. Extensive material analysis was performed to investigate the effects of laser irradiation on Ti/Al/n-GaN. Resistance measurements of the various annealing conditions were examined. Dependence of annealing conditions on the contact formation will also be reported.

INTRODUCTION

Forming low resistance thermally stable and uniform ohmic contacts to wide band gap semiconductors such as gallium nitride (GaN) and related materials are desirable for high performance levels in photonic and electronic device applications.[1-3] Despite significant progress being made in the growth and processing technology of light emitting diodes (LEDs), several technical obstacles remain to be solved to improve the electrical efficiency of devices.

It is well known that the formation of low resistance and thermally stable ohmic contacts are critical for improving device performance, since ohmic contact resistance limits the overall performance of optical and electronic devices. The presence of the insulating native oxide at the metal/semiconductor interface results not only in poor interfacial structures but also in an increase of effective Schottky barrier height to act as a barrier for the carrier to transport from the metal to the semiconductor.[4,5] High temperature thermal annealing during device processing is widely used to remove the native oxide on the GaN surface.[6] However, surface roughness and decomposition are usually induced during a thermal annealing process at high temperature, resulting in poor performance and reliability of the devices due to the nonuniform current flow at the damaged interface.[7] Therefore, this work investigates the effects of laser irradiation on metal-semiconductor layers to lower the specific contact resistance and reduce the oxide formation on GaN for improved contact characteristic.

It is reported that laser annealing decreases the Schottky barrier height for metal contacts on n-type GaN due to the creation of N vacancies through the preferential loss of nitrogen atoms. Whereas, in p-type GaN, laser irradiation could increase the hole concentration of the p-type GaN, leading to a decrease in contact resistivity. KrF excimer laser irradiation was also effective in removing native oxide from a GaN surface, leading to a decrease in contact resistance.[8] KrF excimer laser irradiation treatment of a n-GaN surface has been reported to enhance the activation efficiency of Si dopants,[9] and laser irradiation in an air ambient followed by removal of a gallium oxide layer with subsequent HCl cleaning reduced the contact resistance.[10] However, a detailed examination of the Ohmic contact formation mechanism and its relation to the laser irradiation conditions has not been reported.

In this paper, we adopt a novel process which employs laser annealing for metal contact formation on n-GaN. Ti (35nm)/Al (150nm) contact schemes were ebeam evaporated onto n-GaN. Laser annealing was carried out in purging N_2 ambient using a 248nm pulsed KrF excimer laser. A systematic study of annealing conditions on contact properties was performed. Laser irradiation using multiple (1, 5, 10 & 20) pulses at various laser fluences in the range of 0.18 to 0.7 J cm^{-2} was carried out to study the effects of repeated irradiation on Ti/Al.

EXPERIMENTAL DETAILS

In this work, we have carried out growth of GaN epilayers on silicon-on-insulator (SOI) substrates by metal organic chemical vapor deposition (MOCVD). Ti (35nm) was first evaporated onto the n-GaN surface, followed by Al (150nm). The vacuum condition was maintained between the two metal deposition steps, as shown in Figure 1. This was to reduce any possible oxide formation between the two metals.

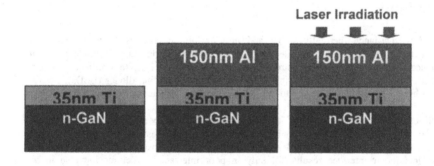

Figure 1. Schematic cross sectional views of Ti and Al metal layers deposited on n-GaN for n-contact fabrication sequence. 35nm Ti was deposited, followed by 150nm Al. After which, the layers were subjected to laser irradiation.

TEM IMAGE OF Ti/Al FORMATION ON GALLIUM NITRIDE

Bright field cross sectional TEM images prepared from Ti/Al contacted samples after undergoing (a) RTP and (b) Laser Annealing, are illustrated in Figure 2. TEM is performed to investigate the morphological changes in Ti/Al. After undergoing laser annealing, the Al and Ti interface remains clearly visible, but when RTP was performed, the Ti/Al interface is unclear. A significant change in the integrity of the contact/nitride interface was observed for samples thermally annealed at or above 880°C. For laser annealed samples, in a broad spectrum of annealing conditions (laser fluences in the range of 0.18 to 0.7 J cm^{-2}), the sample shows an abrupt interface, and laser annealing, no penetration of material further into the GaN/AlGaN was observed.

The interface between the contact electrode materials and semiconductor are not degraded when laser irradiation using multiple (1, 5, 10 & 20) pulses was performed. This shows that the interfaces between the contact materials and GaN are not affected by the short duration, high intensity laser annealing conditions.

TLM STRUCTURES AND I-V MEASUREMENTS OF OHMIC CONTACTS

Room temperature TLM measurements of Al/Ti contact to GaN prior to and after laser annealing are performed. Measurement of current-voltage (I-V) data was obtained using parameter analyzer (HP4155A).

Figure 3 shows the I-V characteristics for nonalloyed Ti/Al contacts before and after laser irradiation of the n-GaN layers. The as-deposited Ti/Al contact sample showed rectifying contact behavior over a range of voltages. However, the laser irradiated sample showed linear I-V behavior, indicating that good ohmic contacts were formed on the n-GaN layers. The specific contact resistance was then determined from plots of the measured resistance versus the spacing between the TLM contacts. The linear square method was used to fit a straight line to the experimental data. To determine the specific contact resistivity of the laser annealed sample, resistances are measured around 0V. The specific resistivity extracted from various thermal and laser annealing conditions. The estimated specific contact resistivities are plotted and summarized in Figure 4.

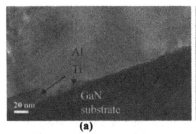 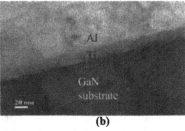

(a) (b)

Figure 2. Bright field TEM images of Al/Ti/GaN layer after undergoing (a) rapid thermal annealing at 880°C, 30sec (b) laser annealing at 0.52 Jcm^{-2}, five pulses. Rapid thermal annealing shows the development of compositional segregation due to the consumption of Ti and diffusion of Al, and laser anneal shows clearly defined interfaces remained, the Ti/Al interface can be discerned.

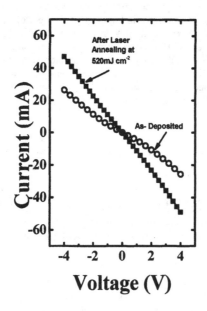

Figure 3. *I-V* characteristics of as deposited Ti/Al contacts and contacts formed after laser annealing at an optimized laser annealing condition of 520mJ cm^{-2} on n type GaN.

The advantage of using laser annealing was that it led to the achievement of a lower contact resistivity.0.52J cm^{-2} 5 pulses laser irradiation reduced the contact resistivity of the film to 4.8e-6ohm-cm as compared to RTA at 880°C,30s [4.75 ×10^{-5} ohm-cm^{2}]. Lower contact resistivity is attributed to minimal reaction of GaN with TiAl during the short duration irradiation. Increasing laser energy from 0.3 J cm^{-2} to 0.52 J cm^{-2} improved contact resistivity from 3.6 ×10^{-6} ohm-cm^{2} to 2.4 ×10^{-7} ohms-cm^{2}. However, contact resistivity is degraded at higher laser energy of 0.7 J cm^{-2} (6.5 ×10^{-6} ohm-cm^{2}). At 0.7 J cm^{-2}, due to a larger melt depth, Ga interaction with Al induced surface roughening and increased sheet resistance.

IMPACT OF LASER ANNEALING ON CONTACT SURFACE MORPHOLOGY

At an optimal laser fluence of 0.52 J cm^{-2} 5 pulses, Ti was completely consumed to form AlTi, Al$_2$Ti and Al$_3$Ti intermetallic phases that exhibit higher melting points and lower diffusivities than Al alone. This is shown in the glancing angle irradiation XRD spectra of laser annealed and RTP annealed Ti/Al contacts in Figure 5. Since Ti (002) has 2θ=38.422°, while Al(111) has 2θ=38.473°, they have very close diffraction angles and thus form a wide peak at 2θ~38°. Laser annealing at 520 mJ cm^{-2} was essential to react Al and Ti to form low resistive Al$_3$Ti and other intermetallic phases for contact resistance reduction. Al$_3$Ti has 2θ=39.252°. This was vital as excessive Al diffusion is responsible for rough surface and higher contact resistivities. Contact resistance lowering was further attributed to increased content of Al$_3$Ti and AlTi formation from the Al-Ti reaction. Repeated irradiation at 0.52 Jcm^{-2} increased contact resistivity (**1 pulse** [2.15 ×10^{-7} ohms-cm^2], **5 pulses** [2.4 ×10^{-7} ohms-cm^2] and **20 pulses** [6 ×10^{-7} ohms-cm^2]), as it similarly promoted deeper Al diffusion as illustrated in the SIMS profile in Figure 5 and extended the Al tail to a larger depth.

The increase in contact resistivity, was possibly attributed to increased Ti-Al reaction with laser irradiation. SIMS data in Figure 5 and 6 further reveal an increased consumption of Al with laser annealing. Also, a steeper elemental profile reveals a more abrupt junction formed when laser irradiation was employed. Thus laser annealing reduced dopants and Ga or Al diffusivities, thus minimized surface roughening issues. The oxygen penetration depth and interaction with GaN was also reduced when laser irradiation was employed as opposed to RTP. The reduction of the insulating native oxide at the metal/semiconductor interface leads to improved interfacial properties and possibly reduced the effective Schottky barrier height for easier carrier transport across the metal–semiconductor interface.

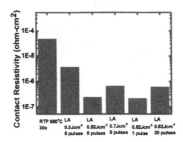

Figure 4. The specific contact resistivity plotted as a function of annealing conditions for Ti/Al contact to *n*-GaN.

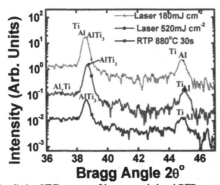

Figure 5. Glancing angle irradiation XRD spectra of laser annealed and RTP annealed Ti/Al contacts. Since Ti (002) has 2θ=38.422°, while Al(111) has 2θ=38.473°, they have very close diffraction angles and thus form a wide peak at 2θ~38°. Laser annealing at 520 mJ cm^{-2} was essential to react Al and Ti to form low resistive Al$_3$Ti and other intermetallic phases for contact resistance reduction. Al$_3$Ti has 2θ=39.252°

83

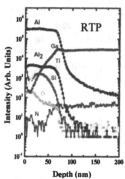

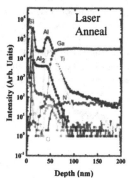

Figure 6. Profiles of elemental concentration versus depth from the surface of the Ti/Al contact after undergoing rapid thermal anneal at 880°C, 30sec.

Figure 7. Profiles of elemental concentration versus depth from the surface of the Ti/Al contact after undergoing laser anneal at 0.52 J cm^{-2}, five pulses.

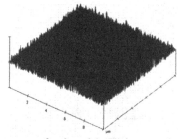

Figure 8. Atomic force microscopy scan of surface of the Ti/Al contact after undergoing laser anneal at 0.52 J cm^{-2}, five pulses. A surface roughness of ~0.5nm was obtained.

Atomic force microscopy (AFM) scan further reveals rms roughness ~0.5nm after laser annealing in Figure 8. Ti/Al on n-GaN with a smooth morphology and improved electrical stability was achieved with laser annealing. With repeated pulse irradiation, or higher laser energy, the rms roughness increased to ~1.5nm. No distinct change in roughness was found after HCl treatment of the laser irradiated surface. AFM results indicate that the laser induced change near the surface region was significant and a large portion of it remained even after the HCl treatment. HCl could remove surface Ga oxide effectively[10], but could not remove the surface roughening caused by the laser irradiation.

CONCLUSION

We investigated the electrical contact formation on n-type GaN film. Low resistance contacts were obtained for n-type GaN using laser anneal. The *I-V* characteristics of the contacts formed were improved from rectifying, in as deposited Ti/Al on GaN, to ohmic, after undergoing laser annealing. In conclusion, contact formation by pulsed laser irradiation is promising for integration with III-V substrates for future device applications.

84

REFERENCES

1. S. Nakamura, M. Senoh, N. Iwasa, and S. Nagahama, Jpn. J. Appl. Phys.,Part 2 **34**, L797 (1995).
2. M. A. Khan, A. R. Bhattarai, J. N. Kuznia, and D. T. Olson, Appl. Phys. Lett. **63**, 1214 (1993).
3. D. L. Hibbard, S. P. Jung, C. Wang, D. Ullery, Y. S. Zhao, H. P. Lee, W.So, and H. Liu, Appl. Phys. Lett. **83**, 311 (2003).
4. J. S. Jang, S. J. Park, and T. Y. Seong, J. Appl. Phys. **88**, 5490 (2000).
5. K. Hattori and Y. Izumi, J. Appl. Phys. **53**, 6906 (1982).
6. M. E. Lin, Z. Ma, F. Y. Huang, Z. F. Fan, L. H. Allen, and H. Mokoc, Appl. Phys. Lett. **64**, 1003 (1994).
7. Y. J. Lin and C. T. Lee, Appl. Phys. Lett. **77**, 3986 (2000).
8. J. S. Jang, S. J. Park, and T.-Y. Seong, J. Vac. Sci. Technol. B **17**, 2667 (1999).
9. D. J. Kim, H. M. Kim, M. G. Han, Y. T. Moon, S. Lee, and S. J. Park, Phys. Status Solidi B **228**, 375 (2001).
10. H. W. Jang, T. Sands, and J. L. Lee, J. Appl. Phys. **94**, 3529 (2003).

85

Mater. Res. Soc. Symp. Proc. Vol. 1252 © 2010 Materials Research Society   1252-I05-20

Micro Probe Carrier Profiling of Ultra-shallow Structures in Germanium

Trudo Clarysse[1], Alain Moussa[1], Brigitte Parmentier[1], Pierre Eyben[1], Bastien Douhard[1], Wilfried Vandervorst[1,2], Peter F. Nielsen[3], Rong Lin[3], Dirch H. Petersen[4], Fei Wang[4], Ole Hansen[4,5]

[1]Imec, Kapeldreef 75, B-3001 Leuven, Belgium
[2]KU Leuven, Dept. of Physics-IKS, Celestijnenlaan 200D, B-3001 Leuven, Belgium
[3]Capres A/S, Scion-DTU, Building 373, DK-2800 Kongens Lyngby, Denmark
[4]DTU Nanotech, Dept. of Micro and Nanotechnology, Technical University of Denmark, Building 345 East, DK-2800 Kongens Lyngby, Denmark
[5]CINF, Center for Individual Nanoparticle Functionality, Technical University of Denmark, DK-2800 Kongens Lyngby, Denmark

ABSTRACT

The performance of electronic devices relies crucially on the precise tailoring of their carrier distributions. The earlier widely used conventional spreading resistance probe (SRP) suffers from many limitations for profiling sub-50 nm silicon based profiles and is also poorly suited for the new high mobility materials being considered today. In this work we therefore investigate into more detail the capabilities of a new approach involving the measurement of the localized sheet resistance along a beveled surface using a micro four-point probe (M4PP) tool (with 1.5 up to 10 μm probe pitch).

INTRODUCTION

For sub-50 nm carrier profiling of silicon based CMOS structures SRP suffers from many issues as discussed in detail elsewhere [1] (need for probe conditioning, large Laplace-based deconvolution correction factors, need for electrical contact radius calibration, pressure induced carrier spilling (10 GPa), lateral upward 3D current flow). In addition, for the new high mobility materials, either probe penetration becomes a limiting factor (20-30 nm on Ge and InGaAs), or contacting problems due to the absence of a β-tin phase transformation (on GaAs).

Recently it has been illustrated that M4PP (with micrometer scale electrode separation) is a very sensitive tool to measure the local sheet resistance along the surface of a silicon wafer [2], and that it is also promising for carrier depth profiling along a beveled surface [3] (no probe conditioning, absolute technique (no calibration needed), virtually zero penetration, no pressure induced carrier spilling, and no β-tin phase needed).

In this work we discuss M4PP sheet resistance depth profiles measured on Ge sub-100 nm structures and evaluate the M4PP intrinsic capabilities relative to SRP.

STRUCTURES

This work focuses on highly B-doped, chemical vapor deposition (CVD) grown, Ge (active concentration of $10^{19}/cm^3$) junction isolated layers with different layer thickness (i.e.

junction depths), in the range 20 to 60 nm depth. In Figure 1 the Secondary Ion Mass Spectrometry (SIMS) dopant profiles for the samples studied are shown. From the SIMS profiles the samples Ge06, Ge08, and Ge10 are seen to have junction depths of approximately 20, 40, and 60 nm, respectively. The underlying structure was 1.5 μm lightly n-type doped Ge (active level of 10^{16}/cm^3) on top of 1.5 μm undoped (slightly p-type) Ge on top of a Si substrate (i.e. p$^+$.n.p$^-$ Ge on Si). All beveled surfaces, as needed to access the in depth information, have been made as for SRP measurements [1]. Angles were about 0.05 up to 0.15 degrees (~500 times magnification).

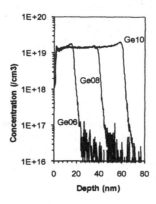

Figure 1: Secondary Ion Mass Spectrometry dopant profiles for Ge B-doped chemical vapour deposited (CVD) layers. Profiles for three samples Ge06, Ge08, and Ge10 with junction depths of approximately 20, 40, and 60 nm are shown.

THEORY

Before discussing the experimental data in to more detail, let us first try to compare SRP and M4PP data from a theoretical point of view.

The simplest way to simulate a M4PP depth profile is to approximate the bevel by a consecutive removal of thin layers of material (of infinite lateral size) and then to calculate the sheet resistance for a known dopant profile over a range of depths (x-direction) up to an insulating boundary. For a box-profile, of main interest here, the sheet resistance at any depth is given by the formula $R_{s,i} = \rho/D_i$, where ρ is the resistivity of the layer and D_i is the thickness of the contacted layer at the i-th depth position (from the boundary). If one sub-divides D_i in i sub-layers with thickness d, one can also write $R_{s,i} = \rho/(i \cdot d)$.

Alternatively, the corresponding SRP profile can be calculated by applying the Laplace multi-layer theory as described in ref. [1]. The latter theory is based on the fact that the measured spreading resistance at a given depth, when contacting a layer with resistivity ρ_i, can be obtained from the formula: $R_i = (\rho_i/2a)^* CF_i(a,s)$, where a is the contact radius and s is the probe separation. CF is a correction factor to compensate for the influence on the current distribution by the layers underneath the contacted one. For those familiar with SRP, note that we neglect any contact resistance or carrier spilling here, as they are not relevant for this discussion. The SRP calculations can be made with different commercially available packages [4,5].

If one now compares these "ideal" SRP spreading resistance (taking a=1 μm, and s=20 μm) and M4PP sheet resistance simulated profiles for sub-100 nm profiles, one will observe that there is virtually no difference between both [1,6]. This is due to the fact that because of its large contact radius and large probe separation, the measured SRP resistance is actually no longer a constriction (spreading) resistance in the real sense of the word (there is no physical space for constriction) but really just the sheet resistance (lateral current spreading) which is measured.

So why should one then theoretically consider M4PP for carrier depth profiling? To answer this question we need to take a closer look at the underlying formulas. Typically both in SRP and M4PP the underlying resistivity (and carrier) profiles are reconstructed by starting at the measurement point closest to the insulating boundary. Next, the subsequent resistivities towards the surface are obtained from the relative changes in measured resistance. For a box-profile (constant $\rho=\rho_i$) in SRP, ρ at each depth is related to $\Delta R_i=R_i-R_{i+1}$ through the formula $\rho=(2 \cdot a \cdot \Delta R_i)/(\Delta CF_i)$. As in SRP correction factors close to an insulating boundary can be as large as a factor 5×10^4, the difference of two large numbers is involved in the denominator, which numerically is ill-advised. On the other hand, in M4PP (also for a box profile) we have $\rho=d \cdot i \cdot (i+1) \cdot \Delta R_{s,i}$, which is numerically much more favorable.

There is a second issue which gives M4PP an advantage over SRP. In practice, any measurement will always have some noise. Despite the application of a good smoothing algorithm [7], some noise will always remain. Hence, the noise propagation of the applied resistivity (carrier) extraction scheme is quite important. In SRP errors have the tendency to build up, as the CF is dependent on the whole underlying profile, which may already have unwanted noise induced variations from earlier calculation steps. In M4PP, as one determines the resistivity of the next (closer to the surface) sub-layer ρ_i solely from the measured sheet resistance of the underlying layer, no errors in underlying resistivity profile shape can be propagated.

Finally, in M4PP we must also take in to account the presence of the bevel surface (the y-direction). It has been illustrated before that M4PP is sensitive to lateral variations in sheet resistance in the immediate neighborhood of the probes (within about one probe pitch distance) [8]. Figure 2a shows a series of 1D simulations (starting from SIMS) for different probe pitches for the Ge10 structure (60 nm junction depth). Clearly the smaller the pitch, the smaller the distortion is expected to be. In earlier SRP work this behavior can be related to the so called upward 3D lateral current flow [1].

GERMANIUM

Figures 2b and 2c show the M4PP sheet resistance profiles, which have been obtained on the Ge CVD layers (Figure 1) in comparison with SRP. The experimental M4PP settings used were: 10 μA current, 11 Hz modulation frequency and 0.4 μm engagement depth. First, note that the assumed junction position of all experimental profiles has been aligned at 60 nm depth, i.e. thickness of the thickest structure, for ease of comparison.

If we compare the M4PP profiles measured at 10 μm (Imec) and 1.5 μm (Capres) pitch, clearly one can observe that the latter ones are much closer to the ideal theoretical behavior. The out smearing of the raw data in the 10 μm pitch case as seen for the 60 nm thick CVD layer(Figure 2c), and for the 20 nm thick CVD layer (Figure 2b), is in qualitative agreement with the theoretical sensitivity behavior (Figure 2a). Clearly it will be quite difficult to extract the underlying carrier profile with high accuracy from the latter raw data. Note that the 10 μm and 1.5 μm profiles (Figure 2b & 2c) for the same structure have been aligned on the expectation that the Rs value at half the top layer thickness is pitch independent (as illustrated in Figure 2a).

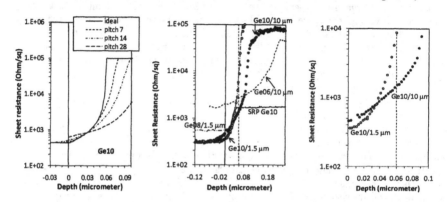

Figure 2: (a) Theoretical impact of the sensitivity of the M4PP probes to their surroundings for sample Ge10 (dependent on probe pitch in micrometer) on the raw bevel data, (b) experimental M4PP Rs profiles with 10 and 1.5 μm pitch on Ge06, Ge08 and Ge10 structures versus SRP, (c) magnification of (b) only displaying the data for Ge10 (dashed line is metallurgical junction).

Note that all raw data converge, beyond the junction, towards a sheet resistance value of about 10^5 Ohm/sq (actual range is 5×10^4 up to 5×10^5 Ohm/sq, dependent on measurement configuration and used sample/bevel). Based on the growth conditions, a sheet resistance for the underlying layer of only 2000 Ohm is expected (cfr. also SRP profile). The much higher observed sheet resistance can, however, easily be understood by the impact of surface states on the beveled surface. The latter are known to create a thin (somewhat irreproducible) p-type inversion layer in lightly doped n-type Ge. Hence, the measured structure rather looks like a $p^{++}p$ structure than a $p^{++}n$ structure to M4PP, similar as for SRP and SSRM [9]. For completeness, it should be mentioned that near the bevel edge part of the rounding of the raw data is probably due to bevel edge rounding (no capping oxide).

Figure 2b also shows the SRP raw data for sample Ge10. The shapes of the SRP and M4PP (1.5 μm pitch) profiles in the top layer are similar as expected. However, note that the apparent dynamic range of the M4PP profile is larger than the SRP one (10^4 ohm vs 2×10^3 ohm). This is as mentioned above because M4PP measures the sheet resistance of the inverted surface layer, while SRP punches through this layer and also senses the underlying n-type material.

CONCLUSIONS

As there is a continuous need for accurate, high resolution, one-dimensional, carrier depth profiling for sub-50 nm structures, both in Si, and new high mobility materials, we have investigated here the theoretical and experimental features of a micro four-point probe (M4PP) profile sheet resistance tool approach along a beveled surface relative to conventional SRP.

Experimentally both SRP and M4PP resistance depth profiles on the sub-100 nm Ge structures considered here are basically identical in the highly doped top layer as expected from theory. As for SRP, one needs an oxide capping layer to accurately determine the profile starting point in M4PP. Similar to SRP, M4PP is also sensitive to geometrical (bevel induced) and surface states induced carrier spilling (and needs to be corrected for it).

On the M4PP pro's side, it needs no probe conditioning, it is an absolute measurement technique (no calibration needed), has virtually zero penetration, does not suffer from pressure induced carrier spilling (no β-tin phase transformation needed), and therefore, works well on Ge. Furthermore, an improved sensitivity to the underlying resistivity profile (extraction more stable numerically) and a reduced noise propagation relative to SRP is expected. The latter is exemplified by the fact that with only very basic smoothing one can already obtain quite useful carrier profiles.

On the M4PP con's side, it is quite important to use probes with a small enough pitch. The pitch should be 1.5 μm or preferably even less to get experimental data comparable to SRP. Otherwise, a deconvolution algorithm to compensate for the impact of the lateral sensitivity of the probes to a region within about one probe pitch from the probe is essential (this is even the case for the 1.5 μm pitch). Probe alignment with the bevel edge is also somewhat tedious in the absence of a capping oxide (lack of visual contrast).

ACKNOWLEDGMENTS

The authors would like to acknowledge Solecon (US) for the spreading resistance probe data and Vincent Benjamin (imec) for the growth of the Ge CVD layers. Also, Ilse Hoflijk (imec) for the beveling of the samples, and Joris Delmotte (imec) for the bevel angle measurements. CINF is sponsored by the Danish National Research Foundation.

REFERENCES

1. T. Clarysse, D. Vanhaeren, I. Hoflijk, W. Vandervorst, Mater. Science & Engineering Reports, 47, 123 (2004)
2. D.H. Petersen, et al., J. Vac. Sci. Technol. B 28, C1C27 (2010)
3. T. Clarysse, P. Eyben, et al., J. Vac. Sci. Technol. B 26, 317 (2008)
4. T. Clarysse, Imecprof, Professional Spreading Resistance Probe analysis package, Imec, Belgium (2003)
5. Solid State Measurements (Semilab), SRP2 package, Pittsburgh, US
6. M. Pawlik, Fourth International Workshop on measurement, characterization and modeling of Ultra-Shallow Doping profiles in semiconductors, p. 26.1 (1997)
7. T. Clarysse, W. Vandervorst, Solid-State Electron. 31, 53 (1988)
8. Fei Wang, et. al., J. Vac. Sci. Technol. B, 28, C1C34 (2010)
9. T. Clarysse, P. Eyben, et. al., J. Vac. Sci. Technol. B 24, 381 (2006)

III-V MOSFET

Mater. Res. Soc. Symp. Proc. Vol. 1252 © 2010 Materials Research Society   1252-I06-04

Band Offset Control by Interfacial Oxygen Content at

GaAs:HfO$_2$ Interfaces

Weichao Wang[1], Robert M. Wallace[1,2] and Kyeongjae Cho[1,2,*]

[1]Department of Materials Science & Engineering and [2]Department of Physics, the University of Texas at Dallas, Richardson, TX 75252

*kjcho@utdallas.edu

Abstract- The impact of interfacial oxygen content on the band offsets of GaAs:HfO$_2$ interfaces was investigated using the density functional theory (DFT) method. Reference potential method was used to determine the band offsets. Moreover, GW correction was utilized to find more accurate value of the valence band edge of HfO$_2$ and hence obtain more accurate band offsets. With gradually decreasing the interfacial O content from 100% to 30% (by changing O chemical potential corresponding to varying the growth condition), the valence band offset increases from 1.06 to 3.34 eV. It is found that this increase of the valence band offsets is inversely proportional to the charge loss of interfacial Ga atoms. Specifically, less charge loss of interfacial Ga induces less charge transfer from GaAs to HfO$_2$ side. Consequently, the less charge loss of interfacial Ga essentially leads to an increase of the valence band offsets.

Introduction - Hafnium oxide (HfO$_2$) has attracted much research interest as high-k gate dielectric material, and it has shown a potential to replace SiO$_2$ for future metal oxide semiconductor devices. It is worthwhile to note that a wide employment of Si as the basic semiconductor was motivated by the excellent properties of its interface with native oxide SiO$_2$. The replacement of SiO$_2$ by a different dielectric (e.g., HfO$_2$) on Si channel has been successfully accomplished in the microelectronics community through the HfO$_2$/Si interface optimization,[1,2] and it is currently followed by an attempt to replace the Si channel with high-mobility substrate, such as GaAs, for further device scaling.[3] Among the stringent interface properties that must be satisfied to achieve optimized device performance, the band alignment at the semiconductor/oxide interface is a key characteristic.

There have been many efforts to study the band offsets (BO) of the oxide/semiconductor interfaces.[4,5,6,7] Robertson[6] has developed a model analysis of the

BOs of many candidate oxides, based on the model of metal-induced gap states (MIGS), charge neutrality level (CNL), and Schottky barrier pinning.[7] However, to understand the realistic interface properties based on the optimized interface atomic configurations, it is also necessary to consider the interface atomic structures and the corresponding electronic structures including interface dipole. Specifically, interface dipole is believed to play an important role on the band alignment and can significantly change the band offset values up to 1 eV in the case of Si-based interfaces. For GaAs:HfO$_2$ interfaces, since the ambient oxygen pressure during the oxide growth can effectively change the interface oxygen content, there should exist a large variation of interfacial dipoles among different interfaces, resulting in a significant variation of the band offsets. Nevertheless, the mechanisms of the variation of band offsets which are induced by different interfacial oxygen content still remain unknown. Therefore, it is necessary to examine the impact of the interfacial oxygen content on the interface band offsets.

In this work, a theoretical study of the electronic properties of GaAs:HfO$_2$ interface is reported. The impact of the interfacial oxygen content on the BO value is discussed in the results and discussion section. A further analysis on the origin of BO's dependence on the interfacial oxygen content is performed as well.

Methodology - Our calculations are based on the density functional theory (DFT) method with the PW91 version of the generalized gradient approximation (GGA) for the exchange-correlation potential, as implemented in a plane-wave basis code VASP.[8,9] The pseudopotential is described by projector-augmented-wave (PAW) method.[10] An energy cutoff of 400 eV and an 8×8×1 k-point with a Gamma centered k mesh were used in our calculations. The force on each atom was converged to 0.01 eV/ Å during the atomic structure optimization.

To build a model interface, we consider the interface between cubic HfO$_2$ and GaAs. Although HfO$_2$ exists in cubic, tetragonal, and monoclinic phases, they all have very similar local ionic bonding, and the atomic structures are closely related.[11] The monoclinic phase with the lowest symmetry is the most stable and is usually detected in deposited films. However, since the cubic phase is the simplest one, it is used in the initial model interface structures. In this work, cubic HfO$_2$ is allowed to change into lower energy structures during the atomic structure optimization. Furthermore, the conclusions from the current analysis would be applicable to other phases because the key requirement is valence satisfaction which depends on the local bonding configuration rather than long-range crystalline symmetry. We use a periodic slab model with Ga-O bonds at the interface (formed by Ga terminated GaAs and O terminated HfO_2 surfaces), which is supported by experimental data.[12,13] A 10 Å vacuum region was used to avoid the interactions between top and bottom atoms in the periodic slab images. The bottom Ga atoms are passivated by pseudo-hydrogen (with 1.25 valence electron) to mimic As-bulk bonds. Meanwhile, the top layer of HfO_2 is initially terminated by 10 oxygen atoms in the unit cell, and half of them are removed to generate an insulating HfO_2 surface without surface states. The applied passivation of the top (HfO_2) and bottom (GaAs) surfaces guarantees that the top and bottom surface states are removed, and all the calculated gap states originate from the interface. The GaAs slab is 27.16 Å thick with 10 layers of Ga and 9 layers of As, while the HfO_2 slab is 13.42 Å thick with 5 layers of Hf and 6 layers of O. We perform a full relaxation by the CG optimization method with only bottom of the passivated GaAs layers fixed.

Results and Discussion - Fig.1 shows the optimized interface model in which there are ten interfacial oxygen atoms, and this interface model is notated as O10. In this work, interfaces are notated by the number of their interfacial oxygen atoms. At the interface O10, the interfacial Ga-As bonds are ~0.170 Å longer than those in the GaAs bulk (~2.472 Å). The movement of the interfacial Ga accompanied by extended Ga-As bonds induces a strain along z-direction on the As layer leading to As movement accordingly. As a consequence, one of the Ga-As bonds between the second (As) and the third atomic layers (Ga) of GaAs is broken, and *the optimized interface structure forms a Ga dangling bond and two As-As dimer pairs*. These interface atomic structure reconstructions have important implications on the interface electronic structures including the band offsets. After removal of one interfacial O atom at different sites in the interface unit cell of the O10 model, the resulting O9 structures are optimized. Different O atoms are removed to generate different O9 models and the lowest energy interface structure is selected for the study. Similarly, more oxygen atoms up to seven are removed to generate different interfaces O8-O3. In the case of O8, it satisfies bonding rules and meets the valence electron saturation requirement. At this specific interface, As-As dimer pairs are broken. The Ga dangling bond is not produced due to a favorable chemical bonding environment. The GaAs atomic structure near the interface is reduced to bulk GaAs structure by removing the reconstructed interface structures present in the O10 and O9 models. With the further decrease of the interfacial oxygen (O7-O3), different interface reconstructions are induced and Ga-Ga dimer pairs, Ga and As dangling bonds begin to occur. Such structures could lead to variable BOs due to the big change of the interfacial oxygen content.

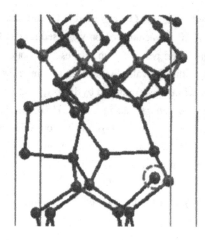

Figure 1 Ball-and stick model represents side view of GaAs:HfO$_2$ interface O10. The Ga, As, Hf and O atoms are depicted by grey, purple, light blue and red balls, respectively. The Ga atoms with dangling bonds at the second top Ga layer are highlighted by green dashed circles.

A schematic diagram of band offset at GaAs:HfO$_2$ interface is shown in Fig. 2. In this work, BOs are determined using the reference potential method.[14,15] First, the top of the VB is found relative to the bulk average potential for pure bulk GaAs and oxide (HfO$_2$). Then the shift in the average potential across the interface from GaAs to the oxide is derived from the calculated electrostatic potentials of the supercell. For atoms far enough from the interface (representing "bulk" atoms), it allows us to determine the difference of the VB edges between GaAs and HfO$_2$. Since the GGA underestimates the band gaps, the CB offsets should be derived by the experimental band gap values of GaAs and HfO$_2$. The GW correction[16] was applied to find the VB edge of HfO$_2$ since GGA could not predict metal oxides VB edge values accurately. The VB edge of GaAs with the GW correction was found to be nearly same as that with regular DFT calculation. Therefore, the GW correction for BO is essentially determined by the VB edge correction of HfO$_2$ rather than GaAs. In the present work, this correction of the VB edge is up to 0.41 eV with respect to standard DFT calculations, and this correction is consistent with what was found previously by Ha, *et al.*.[17]

The calculated VB offsets of various interfaces are summarized in Table I. For the fully O-terminated interface (O10), the VB offset is 1.06 eV after including the GW correction. Since the valence band maximum (VBM) of HfO$_2$ is dominated by oxygen 2p states, the amount of interfacial oxygen has great impact on the VBM of metal oxides but not on that of GaAs. Thus, the VB offsets of the HfO_2/GaAs interfaces are changed significantly by the interface O content. For O9, the VB offset increases by 0.75 eV compared to that of O10. This increase is attributed to the shift down of the HfO_2 VB edge. Table I shows that the VBO kept increasing with decreasing oxygen content at various interfaces.

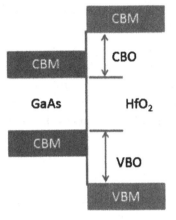

Figure 2 A schematic diagram of band offset at GaAs:HfO_2 interface. VBO and CBO represent valence band offset and conducting band offset between bulk GaAs and HfO_2, respectively.

We now compare these results with other theoretical or experimental results. Robertson has used a semi-empirical CNL method[6] to predict a VB offset of 3.0 eV for GaAs/HfO_2 interface, which corresponds to the VB offset of O6 interface in Table I. Experimentally, for HfO_2 on GaAs, Seguini et al. found a VB offset of 2.10 eV,[18] Dalapati et al. found a VB offset of 2.85 eV,[19] and Afanas'ev et al. found a VB offset of ~2.0 eV,[20] all by photoemission measurements. These values correspond to the O9, O7 and O6 interfaces presented in Table I. This significant variation of the experimental VB offset values may indicate a significant difference in the interface atomic compositions of the experimental samples. Specifically, the sample interface quality depends on the thermal growth condition. For instance, if oxygen pressure is insufficient during the whole growth process (corresponds to Hf rich condition in our previous work[21]), O vacancies would be dominant. The interfacial dipoles vary accordingly based on the content of O vacancies, which directly determines the band offsets of the specific interfaces. O9 interface (VBO = 1.81 eV) is stable over a relatively broad range of μ_O and agrees with the experimental data of 2.10 eV [28] and ~ 2.00 eV [29]. However, neither O7 nor O6 (VBO ~ 2.75-2.98 eV) is the most stable interface in a wide range of μ_o even though they correspond to the experimentally measured VB offsets of 2.85 eV. This difference suggests that the experimental interface formation can be determined by both kinetic process and thermodynamics (determined by O chemical potential) rather than thermodynamics alone as discussed by Robertson *et al.*. [22]

However, as noted in ref. 18, there also appears to be poorly understood differences in the band offsets determined by inverse photoemission[18,20] compared to those obtained from x-ray photoelectron spectroscopy (XPS),[18,19] which may be attributed to dielectric charging issues is XPS experiments. Such effects are important to consider for thick insulators ($\geq$10 nm) such as that reported in ref. 19, but should be minimal for thin ($\leq$3nm) dielectric films, particularly for non-monochromatic laboratory x-ray sources where a significant stray, low energy electron flux is often present from the anode and can neutralize charging effects. Measurements using high intensity x-ray sources, such as synchrotron sources, must also be carefully scrutinized for such effects.[23] Clearly, further systematic studies of such effects is clearly needed for the GaAs system to enable unambiguous conclusions of the source of charging.

Table I. Valence band offsets (in eV) calculated by reference potential method. The GW correction is considered for correcting the GGA error.

Interfaces	GW correction	Ref. pot. method
O10	1.06	0.65
O9	1.81	1.40
O8	2.50	2.09
O7	2.75	2.34
O6	2.98	2.57
O5	3.17	2.76
O4	3.25	2.84
O3	3.34	2.93
CNL[6]	3.00	
Exp.[19]	2.85	

Exp.[24]	2.10
Exp.[20]	2.00

In order to find the impact of interfacial oxygen on the band offsets, Bader charge analysis[25] is performed to find each atom's charge state. Fig. 3 depicts interfacial Ga(As) charge (in e) and VBO (in eV) which share the same scale. Interfacial Ga(As) charges indicate average charges of the first top Ga(As) surface layer in GaAs side. In the case of interfacial As charge, there is little variation among O3-O10 besides O8. This constant As charge suggests that interfacial As dimers are keeping stable electronic configurations under varying interfacial O content (except for O8) and consequently interfacial As has negligible impact on the BO change. For O8, it is a special interface model which satisfies electron accounting rule and the As dimers revert to bulk GaAs bonding. The As shows less charge loss (relative to As atom in bulk GaAs) compared to the rest of interfaces.

Interfacial Ga charge state varies from 1.53 e to 3.03 e compared to 5.40 e ~5.59 e of interfacial As shown in Fig.3. Apparently interfacial Ga charge states and VBO have the same decrease trend versus interfacial oxygen content. The change of the interfacial Ga charge states arise from the variety of electron negativity of the total interfacial oxygen among O3-O10. It means that VBOs decrease when interfacial Ga charge loss decreases according to the increasing of the oxygen content at interfaces. Therefore, less oxygen content at the interface (corresponding to less electron negativity) displays a reduced charge transfer from GaAs to HfO_2 at the interface. The reduced charge transfer of GaAs leads to a smaller shift down of VB edge of bulk GaAs (see Fig. 2). This mechanism explains why the VB offset keeps increasing when decreasing of the interfacial oxygen from O10 to O3. This result indicates that the band offset could be controlled experimentally by changing the interfacial oxygen concentration, which depends on the ambient oxygen concentration.

In reality, the real HfO_2/GaAs interface is more complicated than the present model. However, the Ga-O bonding dominates the interface which reflects the major

HfO$_2$/GaAs interfacial feature based on XPS results[13]. Other trivial interfacial bonding may exist and modify the electronic structure as well. Nevertheless, trivial interfacial bonding could not change the electronic structure qualitatively. As a result, Ga-O bonding interface model is enough to qualitatively describe the VB offsets dependence of interfacial oxygen content

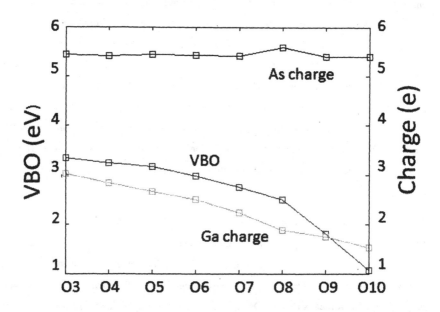

Figure 3. Red square depicts valence band offsets (in eV) by using reference potential methods with GW correction vs. different interfaces (O3-O10). Green and black squares represent average charge (in e) of interfacial Ga and As vs. O3-O10, respectively.

Conclusion - First principles method was used to study the impact of interfacial oxygen contents on the band offsets of GaAs:HfO$_2$ interfaces. We found that by gradually decreasing the interfacial O content from 100% to 30%, the VB offsets can be changed by up to 2.28 eV. Decreasing oxygen content at the interface induces the decreasing of the interfacial Ga charge loss leading to increase of the VB offsets. This finding reveals that band offset has a strong dependence on the interfacial Ga charge values which is determined by interfacial oxygen content.

Acknowledgements - This research is supported by the FUSION/COSAR project, the MSD Focus Center Research Program, a Semiconductor Research Corporation entity and by the NSF under award ECCS-0925844. We thank the III-V materials research groups at UTD for helpful discussions, in particular Prof. Eric Vogel and Dr. Christopher Hinkle. Calculations were performed at the TACC.

References

[1] K. Cho, Computational Materials Science, **23**, 43-47 (2002).

[2] A. Kawamoto, J. Jameson, P. Griffin, K. Cho, and R. Dutton, IEEE Electron Dev. Lett., **22**, 14-16 (2001).

[3] M. M. Frank, G. D. Wilk, D. Staradub, T. Gustafsson, E. Garfunkel, Y. J. Chabal, J. Grazul, and D. A. Muller, Appl. Phys. Lett. **86,** 152904 (2005).

[4] V. V. Afanas' ev, A. Stesmans, M. Passlack, and N. Medendorp, Appl. Phys. Lett. **85**, 597(2004).

[5] V. V. Afanas' ev, A. Stesmans, R. Droopad,M. Passlack, L. F. Edge, and D. G. Schlom, Appl. Phys. Lett. **89**, 092103(2006).

[6] J. Robertson, J. Vac. Sci. Technol. B **18**, 1785(2000)

[7] W. Mönch, Surf. Sci. **21**, 443(1994).

[8] G. Kresse and J. Furthmüller, Comput. Mater. Sci. **6**, 15 (1996).

[9] G. Kresse and J. Furthmüller, Phys. Rev. B **54**, 8245(1996).

[10] P. E. Blochl, Phys. Rev. B **50**, 17953(1994).

[11] J. Zhu and Z. G. Liu, Appl. Phys. A: Mater. Sci. Process. **78**, 741 (2004).

[12] C. L. Hinkle, A. M. Sonnet, E. M. Vogel, S. McDonnell, G. J. Hughes, M. Milojevic, B. Lee, F. S. Aguirre-Tostado, K. Choi, H. C. Kim, J. Kim, and R. M. Wallace, Appl. Phys. Lett. **92**, 071901 (2008).

[13] C. L. Hinkle, M. Milojevic, B. Brennan, A. M. Sonnet, F. S. Aguirre-Tostado, G. J. Hughes, E. M. Vogel, and R. M. Wallace, Appl. Phys. Lett. **94**, 162101 (2009).

[14] C.G. Van de Walle, R. M. Martin, Phys. Rev. B **35**, 8154(1987).

[15] H. M. Al-Allak and S. J. Clark, Phys. Rev. B **63**, 033311 (2001).

[16] M. S. Hybertsen and S. G. Louie, Phys. Rev. B **34**, 5390 - 5413 (1986).

[17] J. Ha, P. C. McIntyre, K. Cho, J. Appl. Phys. **101**, 033706 _2007.

[18] G. Seguini, M. perego, S. spiga, and M. Fanciulli and A. Dimoulas, Appl. Phys. Lett. **91**, 192902 (2007).

[19] G. K Dalapati, H. Oh, S. J. Lee, A. Sridhara, A. See, W. Wong, and D. Chi, Appl. Phys. Lett. **92**, 042120(2008).

[20] V. V. Afanas'ev, M. Badylevich, A. Stesmans, G. Brammertz, A. Delabie, S. Sionke, A.O'Mahony, I. M. Povey, M. E. Pemble, E. O' Connor, P. K. Hurley, and S. B. Newcomb, Appl. Phys. Lett. **93**, 212104 (2008).

[21] W. Wang, K. Xiong, R.M. Wallace, and K. Cho, Impact of Interfacial Oxygen Content on Bonding, Stability, Band offsets and Interface States of $GaAs:HfO_2$ Interfaces, J. App. Phys. (submitted)

[22] P. W. Peacock, K. Xiong, K. Tse, and J. Robertson, Phys. Rev. B **73**, 075328 (2006).

[23] Tanimura et al., Appl. Phys. Lett. vol.96, 162902 (2010).

[24] G. Seguini, M. perego, S. spiga, and M. Fanciulli and A. Dimoulas, Appl. Phys. Lett. **91**, 192902(2007).

[25] G. Henkelman, A. Arnaldsson, and H. Jónsson, Comput. Mater. Sci. **36**, 354 (2006).

Mater. Res. Soc. Symp. Proc. Vol. 1252 © 2010 Materials Research Society   1252-I06-11

Origins for Electron Mobility Improvement in InGaAs MISFETs with (NH4)2S Treatment

°Y. Urabe[1], N. Miyata[1], T. Yasuda[1], H. Ishii[1], T. Itatani[1], H. Yamada[2], N. Fukuhara[2], M. Hata[2], M. Yokoyama[3], M. Takenaka[3], S. Takagi[3]
[1]National Institute of Advanced Industrial Science and Technology (AIST), Tsukuba, Ibaraki 305-8568, Japan
[2]Sumitomo Chemical, Tsukuba, Ibaraki 300-3294, Japan
[3]The University of Tokyo, Bunkyo, Tokyo 113-8656, Japan

ABSTRACT

We investigated the impacts of surface treatment on the electrical and physical characteristics of Al_2O_3/InGaAs interfaces. $(NH_4)_2S$ and NH_4OH treatments were compared to clarify the origin for the higher electron mobility of the MISFET with the $(NH_4)_2S$ treatment. The Auger and photoelectron spectroscopy studies revealed that the $(NH_4)_2S$ treatment reduces the interface Ga-O and As-O bonds and introduces S atoms at the interface. V_{fb} data indicated that the $(NH_4)_2S$ treatment alters the density and/or strength of the interface dipoles, which supposedly mitigate the carrier scattering at the Al_2O_3/InGaAs interface. Interface trap density near the mid gap was rather unaffected by the $(NH_4)_2S$ treatment.

INTRODUCTION

As the scaling of the complementary metal-oxide-semiconductor (CMOS) devices using the conventional planar-type Si channels is approaching to the physical limit, new channel materials have been attracting much attention [1,2]. In particular, the Ge/III-V dual-channel structure integrated on the Si platform has been recognized as a potential solution to sustain performance improvements in the upcoming sub-10 nm gate-length era [3]. There are many papers that report effective ways for controlling the MIS interfaces to reduce the interface states of III-V MIS devices [4-6]. In these studies, the effectiveness of the $(NH_4)_2S$ surface treatment has widely been recognized for the III-V surfaces. The $(NH_4)_2S$ treatment for the GaAs MIS structure was reported by Nannichi et al, [7]. The effects of the $(NH_4)_2S$ treatment are observed as follows: 1) enhancement in photo luminescence intensity which means low surface state density ($<10^{13}$ cm^{-2}), 2) reduction of oxygen atoms on the surface, 3) clear dependence of the Schottky barrier height on the work function of the contact metal. Xu et al, reported that the $(NH_4)_2S$ surface treatment in the Al_2O_3/InGaAs(100) MIS fabrication is effective to improve the device performance [8]. We recently found that this treatment is effective for the InGaAs(111)A surfaces as well [9]. However, the origins for these improvements are not fully understood. In this paper, we investigated the electrical and physical properties of the MIS interfaces fabricated on the $(NH_4)_2S$- and NH_4OH-treated InGaAs(100) surfaces to identify the effect of surface treatment. We carefully compared FET performance, interface chemistry, and interface electrical properties between the two wet treatments.

EXPERIMENT

MIS capacitors were fabricated on $In_{0.53}Ga_{0.47}As(100)$ layers that were 1 μm in thickness with the doping concentration of 3×10^{16} cm^{-3} were heteroepitaxially grown on n-doped InP(100) wafers by metal organic chemical vapor deposition (MOCVD). The NH_4OH treatment is carried out by soaking the samples in NH_4OH solution for 1 min to remove native oxide such as arsenic and gallium oxides from the surface. The $(NH_4)_2S$ treatment is done by soaking the sample in $(NH_4)_2S$ (0.6~1.0 wt-%) for 10 min at room temperature. Each sample was subsequently rinsed in flowing deionized (DI) water followed by drying the surface using in an N_2 blowgun. After the surface treatment, 13 nm-thick Al_2O_3 layers were grown by atomic layer deposition (ALD) using Trimethylaluminium ($Al(CH_3)_3$) and H_2O at 250°C, with $Al(CH_3)_3$–first sequence. The samples underwent the post deposition annealing (PDA) at 400°C in vacuum for 2 min. The MIS capacitors with TaN_x gate electrodes were fabricated as follows. First, a 30-nm-thick TaN_x layer was deposited on the Al_2O_3 surface by using a sputtering method. Second, electrode patterns of 100-nm-thick Au layer were formed on TaN_x by the lift-off process or stencil mask technique. Finally, this Au pattern acted as the mask in the following reactive ion etching (RIE) of TaN_x using SF_6 as the etchant gas. Electrical contact to the InP substrates was made via a thermally evaporated Al/Au. The fabrication process was completed by post metallization annealing in an N_2 flow at 350°C for 90 sec.

RESULTS and DISCUSSION

Figure 1 shows the characteristics of front-gate MISFETs with standard source-gate-drain configuration which were fabricated by the gate last process [9]. Source/Drain region was formed by Si implantation with the energy of 30 keV and a dose density of 2×10^{14} cm^{-2} which was activated by rapid thermal annealing in an N_2 flow at 600°C for 10s. Well-behaved I_d-V_d characteristics were observed for both the NH_4OH and $(NH_4)_2S$-treated MISFETs as shown in Fig. 1(a). The $(NH_4)_2S$-treated device has higher current drivability than the NH_4OH-treated device. The saturation current at high V_g was limited by the S/D resistance. It was 4 and 2.3 Ω for NH_4OH and $(NH_4)_2S$, respectively, as estimated from the plot of total resistance versus channel length (not shown). The $(NH_4)_2S$ treatment effectively reduces the channel resistance. I_d-V_g curves for each device are shown in Fig. 1(b). I_{on}/I_{off} ratios and sub threshold slope (S.S.) were 10^5 and 10^4, 117 and 160 mV/dec for NH_4OH and $(NH_4)_2S$, respectively. These results indicate that $(NH_4)_2S$ treatment is not effective to reduce the interface trap density.

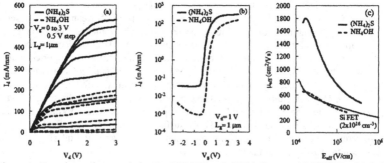

Figure 1. Transistor characteristics for NH_4OH and $(NH_4)_2S$-treated MISFETs. (a) I_d-V_d, (b) I_d-V_g, (c) μ_{eff}-E_{eff}. The MISFETs with channel length (L_g) of 1 μm and channel width of 100 μm were used for the device characteristics (a) and (b). The data (a) and (b) were acquired from the devices with gate length (L_g) of 1 μm and gate width (W) of 100 μm. The data (c) were acquired from the devices with L_g of 150 μm and W of 100 μm.

Figure 1(c) compares the electron channel mobility (μ_{eff}) vs. effective electric field (E_{eff}) for the NH_4OH and $(NH_4)_2S$-treated devices. The μ_{eff} values were extracted from the I_d-V_g, which have the threshold voltage of 0.2 and -0.14 V for the NH_4OH and $(NH_4)_2S$-treated devices, and split C-V data for the channel dimensions of W/L=100/150 µm. The $(NH_4)_2S$-treated device shows high μ_{eff} for the entire E_{eff} range. μ_{eff} at E_{eff}=0.6 MV/cm was 649 cm²/Vs, which is about twice as high as that for NH_4OH and the Si universal mobility. As shown above, $(NH_4)_2S$-treated devices drastically improves μ_{eff}. Now we discuss the possible origins for this improvement.

Figure 2 shows the C-V curves with the frequencies from 100 to 1MHz. The flat band voltage (V_{fb}), which was estimated by using the high frequency (1MHz) capacitance data, is indicated by an arrow in each figure. Both samples exhibit well-behaved C-V curves with small frequency dispersion under the accumulation condition.

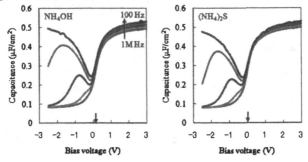

Figure 2. Capacitance-voltage characteristics with frequencies from 100 to 1M Hz for Al_2O_3(12nm)/InGaAs capacitors. The arrows indicate the flat band voltage of V_{fb}.

The G_p/ω map for the $(NH_4)_2S$-treated MIS capacitor is shown in Fig. 3. If the G_p/ω values at the ridge are converted to D_{it} according to D_{it}=2.5 $G_p/\omega/q$, D_{it} was estimated to be ~1x10¹² cm⁻²eV⁻¹.The NH_4OH-treated MIS capacitor showed similar G_p/ω map, and D_{it} was also ~1x10¹² cm⁻²eV⁻¹. Thus, D_{it} is rather unaffected by the choice of the wet treatments.

Figure 4 shows the V_{fb} as a function of oxide thickness. The positive slope of the regression lines indicates fixed negative charges in the oxide or near the oxide/InGaAs interface. Their density is estimated to be of 1.1x10¹² cm⁻² which is independent of the surface treatment. These charges are supposedly generated upon TaN_x sputtering and/or RIE. More importantly, the regression lines intersect with the y-axis at -0.30 V and -0.19 V for the $(NH_4)_2S$ and NH_4OH treatment, respectively. This difference indicates that the density and/or strength of the dipoles at the Al_2O_3/InGaAs interface are varied by the $(NH_4)_2S$ treatment. Existence of sulfur atoms, as shown later in this paper, enhances the dipoles with negative and positive charges on the Al_2O_3 and InGaAs side of the interface, respectively. Such interface dipoles would modulate the microscopic potentials which affect the carrier scattering lifetime. As for the Si MISFETs, the dipole in the high-k gate stack is reported to scatter the carriers in the channel region [10]. Therefore, the changes in the interface dipole by the $(NH_4)_2S$ treatment are thought to be one of the factors contributing to the improved mobility as shown in Fig.1.

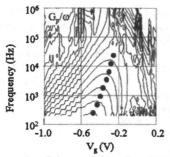

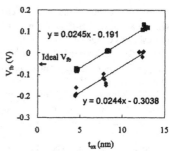

Figure 3. G_p/ω map for the (NH$_4$)$_2$S treatment MIS capacitor at 150 K.

Figure 4. Flat band voltage as a function of oxide thickness for NH$_4$OH (Rectangle) and (NH$_4$)$_2$S treatment (Diamond).

Next, we examine the chemical bonding of the (NH$_4$)$_2$S- and NH$_4$OH-treated samples using Auger electron spectroscopy (AES) and x-ray photoelectron spectroscopy (XPS). Figure 5 shows the AES for a) NH$_4$OH treatment followed by 5 cycles of ALD, b) (NH$_4$)$_2$S treatment surface and c) (NH$_4$)$_2$S treatment followed by ALD. The (NH$_4$)$_2$S treatment surface clearly shows the LMM Auger transition peak of sulfur as shown in Fig. 5(b). After ALD, the S$_{LMM}$ Auger electron peak becomes smaller and, Al$_{LMM}$ and O$_{KLL}$ Auger electron peaks appear in Fig. 5(a) and (c). Moreover, careful analyses of the S $2p$ photoelectron spectra around 162.5 eV in Fig. 6(b) and (c) indicate that the S atoms of about 0.6 monolayer exist at the Al$_2$O$_3$/InGaAs interface. From these results, AES and XPS clearly show that the S atoms are introduced by the (NH$_4$)$_2$S treatment and remain at the Al$_2$O$_3$/InGaAs interface after the ALD.

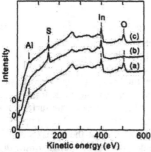

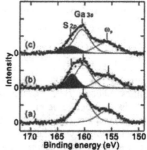

Figure 5. Auger electron spectra of the InGaAs(100) surfaces for (a) NH$_4$OH treatment followed by ALD 5 cycles, (b) (NH$_4$)$_2$S treatment surface, and (c) (NH$_4$)$_2$S treatment followed by ALD 5 cycles.

Figure 6. S $2p$ and Ga $3s$ photoelectron spectra for (a) NH$_4$OH treatment followed by ALD 5 cycles, (b) (NH$_4$)$_2$S treatment surface, and (c) (NH$_4$)$_2$S treatment followed by ALD 5 cycles.

Since we find that the Al$_2$O$_3$/InGaAs interface includes the S atoms, we continue further examination of the XPS data in order to investigate the interface bonding states. Clear chemical shift components of As $2p_{2/3}$ photoelectron spectra can be recognized in Figure 7. It was reported that the chemical shift of 2.9 eV and 3.9 eV respectively correspond to As^{3+} and As^{5+} oxidation states [11]. In addition, the As-S component of about 1.5 eV shift can be recognized for the (NH$_4$)$_2$S treatment. The NH$_4$OH treatment leaves a surface oxide that exhibits a large As oxide peak as shown in Fig. 7. On the other hand, the As oxide peak is reduced by the (NH₄)₂S treatment. While the NH₄OH treatment followed by ALD leaves the higher binding energy As-O peak, (NH₄)₂S treatment followed by ALD almost diminished the As-O and the As-S components.

Figure 7 also shows the Ga $2p_{3/2}$ photoelectron spectra. Initial oxide components are observed for each surface treatment. Since, the chemical shift of Ga-S bond is close to that of Ga-O bond (~1 eV), these components could not be resolved in the Ga $2p_{3/2}$ photoelectron spectra [11]. The Ga-O/Ga-S component for the (NH₄)₂S treatment followed by ALD is smaller than that for the NH₄OH treatment followed by ALD. It is well known that the initial cycles of Al₂O₃ ALD can remove the oxide of the III-V surfaces due to the reducing reaction of Al(CH₃)₃ [12]. The reduction of these Ga-O and Ga-S components indicates that the (NH₄)₂S treatment before ALD assists the cleanup of Ga oxides. The In $3d_{5/2}$ photoelectron spectra also show chemical shift components that may correspond to the In-O and In-S species. It is difficult to deduce the accurate components from the In $3d_{5/2}$ photoelectron spectra, as they are not so surface sensitive compared to the As $2p_{3/2}$ and Ga $2p_{3/2}$ photoelectron spectra. However, we can safely conclude that sub-monolayer of In-O (and Ga-S) components remain at the Al₂O₃/InGaAs interface.

We previously reported that the III-V substrates that form thicker cation- and anion-oxide layers tend to show stronger Fermi-level pinning over the upper half of the band-gap [13]. While this effect contributes to the improvement in the low E_{eff} regime, the S factor in FET was increased by (NH₄)₂S treatment. Therefore, the other factors, such as the dipole as discussed above, should be considered.

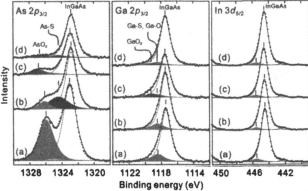

Figure 7. As $2p3/2$, Ga $2p3/2$ and In $3d5/2$ photoelectron spectra. (a) the NH₄OH-treated surface, (b) the S stabilized surface, (c) 5-cycles-ALD on the surface (a), and (d) 5-cycles-

Finally, we address scattering by surface roughness and phonons which are usually considered in the mobility studies. TEM images of the Al₂O₃/InGaAs cross sections were atomically abrupt and smooth for both (NH₄)₂S and NH₄OH treatments. Moreover, RMS roughness measured by AFM was 0.13nm for both surface treatments. Therefore, the difference in the surface and interface smoothness is rather small between these treatments. We also point out that it is unlikely that the probability of phonon scattering changes by the wet treatment of the InGaAs surface.

CONCLUSIONS

We investigated the electrical and physical properties of Al_2O_3/InGaAs interfaces fabricated on $(NH_4)_2S$- and NH_4OH-treated InGaAs(100) surfaces to explore the origin for the mobility improvement by the sulfur treatment. The observed interface chemical bonding and electrical characteristics indicate that the interface Ga-O, As-O bonds is reduced by the $(NH_4)_2S$ treatment. We suggest that interface Ga-S and In-S bonds which modifies the interface dipole in such a manner that the carrier scattering at the Al_2O_3/InGaAs interface is mitigated. D_{it} near the mid gap was rather unaffected by the $(NH_4)_2S$ treatment.

ACKNOWLEDGMENTS

This work was carried out in the Nanoelectronics Project supported by New Energy and Industrial Technology Development Organization (NEDO). A part of this work was conducted at NPF at AIST supported by the Ministry of Education, Culture, Sports, Science and Technology, Japan (MEXT).

REFERENCES

1. Jiezhi Chen, Takura Saraya, Kousuke Miyaji, Ken Shimizu and Toshiro Hiramoto: *Jpn. J. Appl. Phys.* **48** 011205 (2009)
2. Noriyuki Taoka, Masatomi Harada, Yoshimi Yamashita, Toyoji Yamamoto, Naoharu Sugiyama, and Shin-ichi Takagi: *APL* **92**, 113511 (2008).
3. S. Takagi, T. Irisawa, T. Tezuka, T. Numata, S. Nakaharai, N, Hirashita, Y. Moriyama, K. Usuda, E. Toyoda, S. Dissanayake, M. Shichijo, R. Nakane, S. Sugahara, M. Takenaka, and N. Sugiyama: *IEEE Trans. Electron Device* **55** 21 (2008).
4. Han-Chung Lin, Wei-E. Wang, Guy Brammertz, Marc Meuris, Marc Heyns: *Microelectronics. Eng.* 86, 1554-1557 (2009).
5. T. D. Lin, H. C. Chiu, P. Chang, L. T. Tung, C. P. Chen, M. Hong, J. Kwo, W. Tsai and Y. C. Wang: *APL* **93**, 033516(2008).
6. InJo Ok, H. Kim, M. Zhang, F. Zhu, S. Park, J. Yum, H. Zhao, Domingo Garcia, Prashant Majhi, N. Goel, W. Tsai, C. K. Gaspe, M. B. Santos and Jack C. Lee: *APL* **92** 202903(2008).
7. J. Fan, H. Oigawa and Y. Nannichi, *Jpn. J. Appl. Phys.* **27**, L1331-L1333 (1988)
8. Y. Xuan, Y. Q. Wu, T. Shen, T. Yang and P. D. Ye: *IEEE IEDM* 637 (2007).
9. H. Ishii, N. Miyata, Y. Urabe, T. Itatani, T. Yasuda, H. Yamada, N. Fukuhara, M, Hata, M. Deura, M. Sugiyama, M. Takenaka and S. Takagi: *Appl. Phys. Exp* **2**, 121101 (2009).
10. Hiroyuki Ota, Akito Hirano, Yukimune Watanabe, Naoki Yasuda, Kunihiko Iwamoto, Koji Akiyama, Kenji Okada, Shinji Migita, Toshihide Nabatame, and Akira Toriumi: *IEEE IEDM* (2007).
11. S. Arabasz, E. Bergignat, G. Hollinger and J. Szuber: *Appl. Surf. Sci.* 252, 7659 (2006)
12. C. L. Hinkle, A. M. Sonnet, E. M. Vogel, S. McDonnell, G. J. Hughes, M. Milojevic, B. Lee, F. S. Aguirre-Tostado, K. J. Choi, H. C. Kim, J. Kim and R. M. Wallace: *APL* 92, 071901 (2008)
13. T. Yasuda *et al,*: presented at MRS Fall Meeting 2009, Symposium A (proceeding manuscript submitted for electronic publication).

Novel Devices and III-V MOSFET

Mater. Res. Soc. Symp. Proc. Vol. 1252 © 2010 Materials Research Society   1252-I07-03

Study of Germanium Epitaxial Recrystallization on Bulk-Si Substrates

Byron Ho, Reinaldo Vega, Tsu-Jae King-Liu
Dept. of EECS, University of California at Berkeley,
Berkeley, CA 94720-1770, U.S.A.

ABSTRACT

LPCVD Ge films are deposited onto bulk Si substrates and subjected to either a rapid thermal anneal (RTA) or furnace anneal (FA) at a temperature that is higher than the melting point of Ge in an attempt to induce epitaxial recrystallization. Spiking into the Si and voids in the Ge film are observed after the anneal. This is attributed to defect-assisted Ge diffusion into the Si substrate caused by strain at the Ge-Si interface. Simple diffusion theory using published diffusivity values predicts diffusion depths similar to the spiking depths observed by scanning electron microscopy and transmission electron microscopy. Approaches to reduce the strain at the interface are explored. It is found that the quasi-equilibrium nature of FA reduces spiking and that there is an area dependence. Grazing-incidence x-ray diffraction analysis suggests that this technique for epitaxial recrystallization does not result in single-crystalline Ge.

INTRODUCTION

Continued CMOS technology scaling has required the incorporation of advanced materials to sustain the historical pace of improvement in transistor performance. In this context, silicon-germanium ($Si_{1-x}Ge_x$) embedded in the source and drain (S/D) regions of p-channel MOSFETs is now used to compressively strain the channel [1] and thereby enhance the hole mobility. Due to its higher carrier mobilities, Ge is also being considered as an alternative channel material for use in future high-performance integrated circuit applications. To date, Ge MOSFETs have been demonstrated on $Si_{1-x}Ge_x$ substrates and Ge-on-insulator (GOI) substrates [2-4]. Epitaxial Ge-on-Si growth processes using molecular beam epitaxy or ultra-high-vacuum chemical vapor deposition (UHV-CVD) have been developed [5, 6], but these are relatively expensive. Recently, it has been shown that molten Ge can recrystallize epitaxially upon solidification on Si and may provide for a lower cost approach to fabricating Ge MOSFETs on Si or $Si_{1-x}Ge_x$ S/D stressors [7]. There have been several investigations of this approach for silicon-on-insulator (SOI) substrates [8, 9]. In this work, the formation of epitaxial Ge on Si bulk substrates via melting and recrystallization is investigated, to assess the viability of this approach for forming S/D stressors and high-mobility channel regions.

EXPERIMENTAL DETAILS

In this study, Ge was selectively deposited onto (100) Si substrates in an LPCVD furnace, in a series of experimental splits. In Exp. 1, silicon dioxide was thermally grown on the Si substrates and patterned into 1um to 5um wide lines. Then, Ge films were selectively deposited onto the exposed Si regions at 425°C (400mT, 50sccm GeH$_4$). A capping layer of low-temperature silicon dioxide (LTO) was then deposited to prevent evaporation of Ge during annealing. These samples were subjected to a rapid thermal anneal (RTA) for 10sec at temperatures ranging from 850°C to 1000°C, with heating and cooling rates of 50°C/sec and

40°C/sec, respectively. Based on the results of Exp. 1, the Ge deposition temperature was reduced in Exp. 2 to 300°C (800mT, 20sccm GeH$_4$). The films were capped with LTO and annealed with the same RTA conditions as in Exp. 1. For Exp. 3, a thin amorphous Si (a-Si) layer was deposited prior to blanket Ge film deposition. The a-Si (Ge) was deposited at 300°C, 800mT for 3 hours (1 hour) with 100sccm SiH$_4$, 50sccm BCl$_3$ (20sccm GeH$_4$). Due to the low deposition rate, the thickness of the a-Si layer cannot be measured accurately, but it is estimated to be less than 5nm thick. After LTO capping, these samples are annealed either by RTA as in Exp. 1 and 2 or by furnace annealing (FA) for 1hr at 1000°C.

The effect of Ge area was studied in Exp. 4. A 300nm-thick layer of SiO$_2$ was thermally grown, then a buffered HF wet etch solution was used to form large windows (100um)^2, small windows ((24um)^2 – (8um)^2), and narrow lines (3um – 1um wide) in this oxide layer. An arsenic pre-amorphization implant was performed (1×10^{15} cm^{-2}, 10keV) before selectively depositing Ge onto the exposed Si regions and an LTO capping layer using the deposition conditions of Exp. 2. These samples were also subjected to either an RTA or a FA (1000°C, 1hr).

The 425°C-deposited Ge samples were characterized using scanning electron microscopy (SEM) and transmission electron microscopy (TEM). For the 300°C-deposited Ge samples, the LTO capping layer was removed using a wet HF dip and the surface topography scanned in tapping mode using a Veeco Dimension 3100 atomic force microscope (AFM) with a tip radius smaller than 10nm and a tip aspect ratio of 3:1.

DISCUSSION

Experiment 1: Ge deposited at 425°C

The cross-sectional SEM images in Figures 1a and 1b of samples before and after annealing at 1000°C, respectively, clearly show Ge spiking into the Si substrate after the anneal. The depth of the spiking is on the order of 100nm (for an as-deposited Ge film thickness of 70nm), and the Si$_{1-x}$Ge$_x$ surface is pitted. Cross-sectional TEM images of the sample before RTA (Figures 1c and 1d) show that the Ge film is polycrystalline with large grains (60-80 nm), causing non-uniform strain at the Si-Ge interface. During annealing, the strain present at this interface may relax suddenly, causing defects to form. These defects may assist and accelerate Ge diffusion into Si, resulting in the observed spiking.

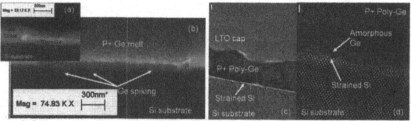

Figure 1: Images of samples with Ge deposited onto Si by LPCVD at 425°C: (a) Cross-sectional SEM image of the Ge film as deposited, (b) Cross-sectional SEM image of the Ge film after annealing, (c) Cross-sectional TEM of as-deposited Ge film, which is polycrystalline, and (d) Close-up of the Ge-Si interface of (c), showing a thin (1-2 nm) amorphous Ge layer at the interface, with strain present in the Si.

Experiment 2: Ge deposited at 300°C

Decreasing the deposition temperature should reduce the average grain size and lower the strain at the Ge-Si interface. AFM measurements show that the rms roughness of the as-deposited film (120nm thick) is 1.4nm compared to 18.8nm for the film deposited at 425°C; this is a reasonable indication that the average grain size of the 300°C deposited film is smaller.

However, the AFM scan of the sample after annealing still shows severe Ge spiking, with a spike density of $\sim 2 \times 10^8$ cm^{-2}. The lateral extent of the spikes varies from 40nm to 500nm, with a maximum spike depth around 100nm for the widest spikes (Figures 2a and 2b). Interestingly, spiking is also observed in samples subjected to an 850°C anneal, which suggests that the spiking is most likely the result of enhanced Ge diffusion rather than a melting phenomenon.

Previous work has shown that Ge diffusion in Si grain boundaries (GBs) is ~ 6 decades faster than in bulk Si, demonstrating the sensitivity of Ge diffusion to crystalline defects [10]. Using published diffusion parameters for GBs in films with 40nm (nano-GB) and 30um (micro-GB) wide grains [10], the Dt product and hence the approximate junction depth can be calculated for the RTA heating conditions used in this study (10.7sec, at 50°C/sec from 400°C to 935°C). From past experience with TEM and energy dispersive x-ray spectroscopy (EDX) measurements, the contrast between Si$_{1-x}$Ge$_x$ and Si disappears for Ge concentrations below $\sim 70\%$. The calculated junction depths (for Ge concentrations above 70%) for the nano-GB (micro-GB) materials are 102.5nm (122.2nm), which agree well with the observed spiking depths in SEM and TEM for Exp. 1. The general agreement between the calculated and observed junction depths as well as spiking observed for sub-melt annealing temperatures imply that the spiking phenomenon is due to accelerated Ge diffusion through defects caused by strain relaxation at the Ge-Si interface.

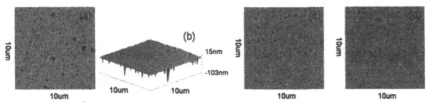

Figure 2: (10um)^2 AFM topography scans of annealed samples: (a) Exp. 2 (RTA 10s, 940°C), (b) 3D isometric plot of (a), (c) Exp. 3 (RTA 10s, 940°C), and (d) Exp. 3 (FA 1hr, 1000C).

Experiment 3: Ge deposited at 300°C with a-Si interlayer

In an attempt to further relieve the strain at the Ge-Si interface, an a-Si interfacial layer is deposited prior to the Ge layer resulting in an rms roughness of 0.6nm for the as-deposited Ge. Spiking is still observed after RTA, however, with a spike density of $\sim 2 \times 10^8$ cm^{-2} (Figure 2c). AFM image software was used to isolate and characterize the spikes from the AFM data, extracting spike depths and planview spike areas. Figure 3b compares the cumulative spike area distribution plots for the various experimental splits. Exp. 3(RTA) results in a larger percentage of smaller-area spikes than Exp. 2(RTA). Figure 3c shows that the total spike area for Exp. 3(RTA) is 47% less than for Exp. 2(RTA), suggesting that the a-Si interlayer helps to mitigate Ge spiking during an RTA melt process. A thicker a-Si layer should further decrease the stress at the Ge-Si interface and thereby reduce the spike depth and area.

For samples furnace annealed at 1000°C for 1hr, the rms surface roughness is 0.9nm, and no spiking is evident. However, AFM scans show large grain boundaries (Figure 2d), which may be due to uneven cooling rates across the substrate resulting in the collision of solidification fronts during recrystallization. The absence of spiking in the FA sample may be a result of the slower temperature ramp of FA compared to RTA, allowing any metastable strain at the Ge-Si interface to relax slowly, minimizing the formation of defects. During the FA ramp, Ge is also diffusing into the Si substrate, creating a graded $Si_{1-x}Ge_x$ layer that should also reduce interface strain. Ge diffusion would be minimal during the short ramp time of the RTA.

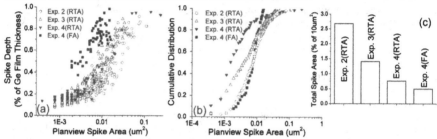

Figure 3: Comparison of spiking in Exps. 2-4: (a) Spike depth vs. planview spike area, (b) Cumulative distribution of spike area, and (c) total spike area for the various experimental splits.

Grazing-incidence x-ray diffraction (GIXRD) analysis results are shown in Fig. 4. Images of the area detector with the diffraction pattern of the Ge layer for as-deposited, 940°C RTA, and 1000°C FA samples are shown in Figures 4a, 4b, and 4c, respectively. The bottom of the hemisphere denotes the top surface of the sample. All annealed films are polycrystalline, but the FA sample has more grain orientations as evidenced by the overlap of discrete diffraction spots into a diffuse ring. Interestingly, a (100) associated peak is not observed in any of the films. Further analysis is needed to understand film orientation after annealing.

Figure 4: GIXRD images of Exp. 3 samples (a) as-deposited, (b) after RTA (10s, 940°C), (c) after FA (1hr, 1000°C).

Experiment 4: Ge deposited at 300°C after preamorphization implant

AFM scans show an rms roughness of 1.8nm for the as-deposited film and spiking in both RTA and FA samples for (100um)² windows. The spike densities are around $1x10^8$ cm⁻² ($7x10^7$ cm⁻²) for the RTA (FA) samples, which is only slightly lower than for the samples in the other experimental splits. Figures 3a and 3b show that for Exp. 4(RTA), most spikes are shallower and smaller than for Exp. 2 and 3. For Exp. 4(FA), the spike area has a smaller distribution, indicating better uniformity and stability of the spiking mechanism. In Figure 3c it can be seen that the total spike area is reduced for Exp. 4. These results indicate that the preamorphization implant achieves an effect similar to that of an a-Si interfacial layer.

AFM scans of RTA and FA samples of Exp. 4 were also performed for oxide window areas of $(24um)^2$, $(16um)^2$, and $(8um)^2$, as well as for the 3um- and 1um-wide lines. The RTA sample shows no area dependence on spiking behavior (Fig. 5a). However, the FA sample shows reduced spike depth and area as the window area decreases to $(8um)^2$ (Figures 5b and 5c). For the 3um- and 1um-wide lines of the FA sample, no spiking is observed and hence there are no corresponding curves in these figures. Note that although a patterned substrate was also used in Exp.1, the film was too rough to provide useful AFM data.

The observed window area dependence on spiking for the FA sample also can be explained by strain effects; in this case, the strain is exerted on the Si wafer by the oxide layer. The buffered HF etch used to open oxide windows on the wafer frontside also removed the thermally grown oxide on the wafer backside. The measured stress in the thermally grown oxide film is around -275 MPa (compressive stress). Using a two-dimensional process simulator [11], a stress value of -250 MPa was assigned to an oxide layer formed on a Si wafer, and an isotropic oxide etch was simulated to open up windows in the oxide layer. From the simulations, the Si surface that is exposed is under tensile stress that generally increases as the window size is reduced (Figure 6). This allows Ge, with a larger lattice constant than Si, to be better lattice-matched to the substrate during deposition and thus reduces the strain at the Ge-Si interface. For RTA, the highly non-equilibrium heating process still causes this decreased strain to relax quickly and form defects, but the more stable FA is shown to benefit from this better lattice matched interface.

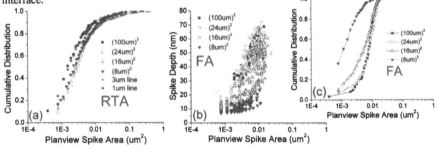

Figure 5: Spike distributions for various Ge areas in Exp. 4: (a) cumulative distribution of spike area for 940°C RTA, (b) spike depth vs. spike area for 1000°C FA, (c) cumulative distribution of spike area for 1000°C FA.

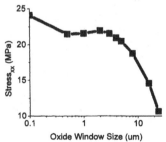

Figure 6: Process simulation results showing stress at the Si surface at the center of an oxide window opening, for various window sizes.

CONCLUSION

LPCVD Ge deposited onto bulk Si and heated to temperatures above the Ge melting temperature result in a spiked Ge-Si interface. Analysis of the experimental results suggest that this spiking phenomenon is due to strain-induced defect-assisted Ge diffusion. An amorphous silicon interfacial layer mitigates the spiking. A slow temperature ramp rate (as in furnace annealing), and process-induced tensile stress help to mitigate strain relaxation and defect formation. Based on the GIXRD analysis results, this approach to forming epitaxial Ge on Si may not be suitable for making high-quality MOSFET channels, but could be a viable alternative to forming embedded $Si_{1-x}Ge_x$ S/D stressors.

ACKNOWLEDGEMENTS

The authors gratefully acknowledge Prof. Eugene Haller, Swanee Shin, and Christopher Liao for TEM analysis and useful discussions. With the aid of Steven Volkman, GIXRD analysis was carried out at the Stanford Synchrotron Radiation Lightsource, a national user facility operated by Stanford University on behalf of the U.S. Department of Energy, Office of Basic Energy Sciences. This project was funded by a Semiconductor Research Corporation GRC Fellowship. Samples were prepared in the UC Berkeley Microlab.

REFERENCES

1. S.E. Thompson, M. Armstrong, C. Auth, S. Cea, R. Chau, G. Glass, T. Hoffman, J. Klaus, Z. Ma, B. McIntyre, A. Murthy, B. Obradovic, L. Shifren, S. Sivakumar, S. Tyagi, T. Ghani, K. Mistry, M. Bhor, and Y. El-Mansy *IEEE Elec. Dev. Lett.* **25**, 191 (2004).
2. G. Hellings, J. Mitard, G. Eneman, B. De Jaeger, D.P. Brunco, D. Shamiryan, T. Vandeweyer, M. Meuris, M.M. Heyns, K. De Meyer, *IEEE Elec. Dev. Lett.* **30**, 88 (2009).
3. G. Nicholas, B. De Jaeger, D.P. Brunco, P. Zimmerman, G. Eneman, K. Martens, M. Meuris, and M.M. Heyns, *IEEE Trans. Elec. Dev.* **54**, 9 (2007).
4. K. Romanjek, L. Hutin, C. Le Royer, A. Pouydebasque, M.-A. Jaud, C. Tabone, E. Augendre, L. Sanchez, J.M. Hartmann, H. Grampiex, V. Mazzochi, S. Soliveres, R.Truche, L. Clavelier, P. Scheiblin, X. Garros, G. Reimbold, M. Vinet, F. Boulanger, and S. Deleonibus, *Solid State Elec.*, **53**, 723 (2009).
5. T. Mack, T. Hackbarth, U. Seiler, H.J. Herzog, H. von Kanel, M. Kummer, J. Ramm, and R. Sauer, *Mat. Sci. Eng. B*, **89**, 368 (2002).
6. M.L. Lee, C.W. Leitz, Z. Cheng, A.J. Pitera, T. Langdo, M.T. Currie, G. Taraschi, E.A. Fitzgerald, and D.A. Antoniadis, *Appl. Phys. Lett.*, **79**, 20 (2001).
7. N. Sugii, S. Yamaguchi, and K. Washio, *J. Vac. Sci. Technol. B*, **20**, 1891 (2002).
8. Y. Liu, M.D. Deal, J.D. Plummer, *Appl. Phys. Lett.*, **84**, 2563 (2004).
9. T.Y. Liow, K.M. Tan, R.T.P. Lee, M. Zhu, B.L.H. Tan, G.S. Samudra, N. Balasubramanian, and Y.C. Yeo, *Symp. VLSI Tech.*, (2008).
10. A. Portavoce, G. Chai, L. Chow, and J. Bernardini, *J. Appl. Phys.*, **104**, 104910 (2008).
11. *Sentaurus Process User Guide*, Version C-2009.06 (Synopsys Inc., 2009)

Mater. Res. Soc. Symp. Proc. Vol. 1252 © 2010 Materials Research Society   1252-I07-05

Interfacial Properties, Surface Morphology and Thermal Stability of Epitaxial GaAs on Ge Substrates with High-k Dielectric for Advanced CMOS Applications

A. Kumar[1,2], G.K. Dalapati[1], Terence Kin Shun Wong[2], M.K. Kumar[1], C.K. Chia[1], H. Gao[1], B.Z. Wang[1], A.S. Wong[1], D.Z. Chi[1]

[1] Institute of Materials Research and Engineering, A*STAR (Agency for Science, Technology and Research), 3 Research Link, Singapore 117602
[2] School of Electrical and Electronic Engineering, Nanyang Technological University, Nanyang Avenue, Singapore 639798

ABSTRACT

Epitaxial GaAs layers had been grown by metal organic chemical vapor deposition at $620^\circ C$ on Ge(100) substrates. The surface roughness of the GaAs is greater than that of GaAs bulk wafers and epilayer morphology is influenced by miscut of the Ge substrate. The GaAs/Ge interface is of good quality and devoid of misfit dislocations and antisite defects. However, Ge diffusion into GaAs occurred during epitaxy and resulted in auto-doping. ZrO_2 was deposited by magnetron sputtering onto the epi-GaAs. Capacitance voltage measurements show that the TaN/ZrO_2/epi-GaAs capacitor has an interfacial layer with more defects than a ZrO_2/bulk GaAs interface. An improved interface with smaller frequency dispersion can be formed by atomic layer deposition of the high-k dielectric layer onto the epi-GaAs.

INTRODUCTION

As the semiconductor industry advances toward the 22nm node for complementary metal oxide semiconductor (CMOS) technology, different new materials are being introduced at both the front end and back end of the line to sustain integrated circuit performance gain [1]. For the channel region between the source and drain, high mobility semiconductors are being intensely investigated as replacement for conventional silicon channels [2]. This is due to the scaling constraints of current strained silicon processes [3]. For n-channel CMOS devices, III-V compound semiconductors such as $In_xGa_{1-x}As$ are being actively investigated for future transistors [4,5]. This is because III-V semiconductors have up to 6 times the bulk electron mobility of silicon [2] and III-V devices can sustain higher junction breakdown voltages. However, it remains challenging to fabricate inversion channel devices because of the difficulties in surface passivation, inadequate understanding of defect states and the need to fabricate high quality high-k III-V interfaces that are free from Fermi level pinning [6].

In this paper, we study the use of epitaxial GaAs grown on Ge substrates as a potential III-V channel material. Hitherto, only bulk III-V wafers have been utilized for III-V nMOSFETs [7,8]. The reason for using Ge susbtrates is that the equilibrium lattice parameters of GaAs and Ge are 0.56633nm and 0.56579nm respectively [9] and their thermal conductivities are comparable. Also, Ge has a high bulk hole mobility of $1900cm^2V^{-1}s^{-1}$ at 300K [2]. Thus, by implementing selective epitaxy of GaAs on Ge, it should be possible to realize integrated CMOS devices based on these two high mobility materials. Methods to fabricate epitaxial Ge layers on Si wafers have been reported [10].

EXPERIMENT

GaAs epitaxy was performed on Ge (100) substrates with a 6° miscut using metalorganic vapor phase deposition (MOCVD) in an Aixtron reactor. Prior to growth, the Ge substrates were heated up at 650°C for 5 minutes in a H_2 ambient in order to desorb the GeO_x native oxides. GaAs layers (260nm) were grown at 620°C by using trimethyl gallium (TMG) and tertiary butylarsine (TBA) precursors. The epitaxial GaAs discussed in this paper are undoped. However, from capacitance voltage (C-V) measurements, the expected doping concentration is in the range of $1-5 \times 10^{16} cm^{-3}$ depending on Ge autodoping. Extrinsic layers were also grown with Zn and Si used as p and n-type dopants respectively. The growth temperature is below the typical optimum growth temperature for GaAs of 650°C. The growth rate of the GaAs was measured to be 0.32nm/s.

After growth, the GaAs epilayer was cleaned by HF for 3 minutes, NH_4OH for 10 minutes and finally rinsed with deionized water. Then a ZrO_2 layer was deposited in a Denton radio frequency (RF) magnetron sputtering system. During deposition, the ZrO_2 target was sputtered using Ar at a RF power of 60W and a pressure of 0.4Pa. Another sample of HfO_2 was deposited by atomic layer deposition (ALD) at 90°C using a home built ALD system with tetrakis (dimethylamino) hafnium (IV) and H_2O as precursors. The deposition rate was determined by ellipsometry to be 0.132nm/cycle. Both samples were rapid thermal annealed in N_2 at 400°C and 700°C after deposition. Finally, TaN gate contacts and Au back contact were deposited by sputtering to form metal-insulator-semiconductor (MIS) capacitor samples.

The surface morphology of the GaAs epilayers was imaged using a Molecular Imaging atomic force microscope (AFM) in tapping mode. The interfaces of the gate stacks and defects within the epilayers were characterized using a JEOL 2100 transmission electron microscope (TEM) with a field emission source and beam voltage of 200kV. Compositional profiles of the sample interfaces before and after post deposition annealing (PDA) was measured using a time of flight secondary ion mass spectrometer (SIMS) at a pressure below $10^{-4}Pa$. C-V curves of the MIS capacitors were measured using an impedance analyzer at frequencies ranging from 10kHz to 100kHz.

DISCUSSION

Surface morphology and TEM analysis

Figure 1a and 1b shows the surface morphology of the GaAs epilayer grown by MOCVD at 620°C on (100) Ge. The scanned area of figure 1a is 10µmx10µm. At this scale, the surface consists of a large number of triangular islands with their axes aligned along a common diagonal. The root mean square (rms) surface roughness of 5.5nm is much greater than the 0.28nm rms roughness of a bulk GaAs wafer with the same scanned area (figure 1c). Figure 1b shows the GaAs morphology over 200nm x 200nm. The surface is rippled with a rms surface roughness of 0.31nm. The GaAs epilayer morphology is influenced by the 6° miscut in the Ge substrate. As a result of this, atomic terraces would be present on the Ge surface and preferential attachment of Ga and As adatoms during growth can occur. The morphology of Zn doped (p-type) GaAs grown on Ge was also characterized. The rms roughness increased with the dopant concentration and the surface appeared granular.

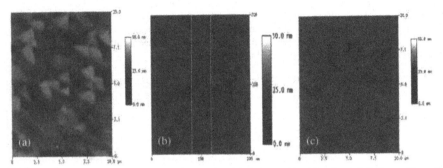

Figure 1. AFM topographic images of: (a) epitaxial GaAs on Ge, (b) epitaxial GaAs on Ge and (c) bulk GaAs wafer.

Figure 2a shows the cross sectional TEM image of a 260nm thick GaAs epilayer on Ge. Although the interface between the GaAs and Ge is abrupt and misfit dislocations are not visible, a defect site indicated by the arrow can be seen near the interface. This could be an antiphase defect but such defects are not found at the epilayer surface. Figure 2b shows a high resolution HRTEM image of the same GaAs/Ge sample with the interface highlighted by a dotted line. The darker areas near the GaAs/Ge interface could be due to defects. However, overall the interface quality is good.

Figure 2. (a) Cross section TEM image of GaAs epilayer grown by MOCVD on Ge at 620°C, (b) high resolution TEM image of epi-GaAs/Ge interface.

Compositional profiles

The elemental compositional profiles of the epi-GaAs/Ge grown at 620°C was measured by dynamic SIMS to a depth of 450nm from the sample surface (figure 3). Consistent with the TEM image, there is an abrupt transition from GaAs to Ge at about 260nm. On the other hand, out-diffusion of Ge into GaAs can also be seen in figure 3. Ge diffused up to 60nm into GaAs from the heterointerface. Since Ge is a group IV element, it can act as an inadvertent n-type dopant. In separate experiments, we have confirmed Ge auto-doping by C-V measurements. A nominally undoped GaAs epilayer can yield a C-V characteristic at high frequency that resembles a n-type semiconductor.

SIMS depth profiles were measured after PDA at 400°C and 700°C. As shown in figure 4, there is a noticeable diffusion of Ga into Ge at 700°C. However, the profiles for both As and Ge changed little during the PDA. The absence of further Ge diffusion during PDA shows that the composition of the GaAs epilayer should be stable during subsequent thermal cycling. This is favorable to device fabrication on the GaAs epilayer.

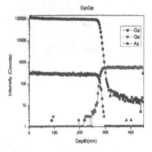

Figure 3. SIMS depth profile of epi-GaAs on Ge after MOCVD.

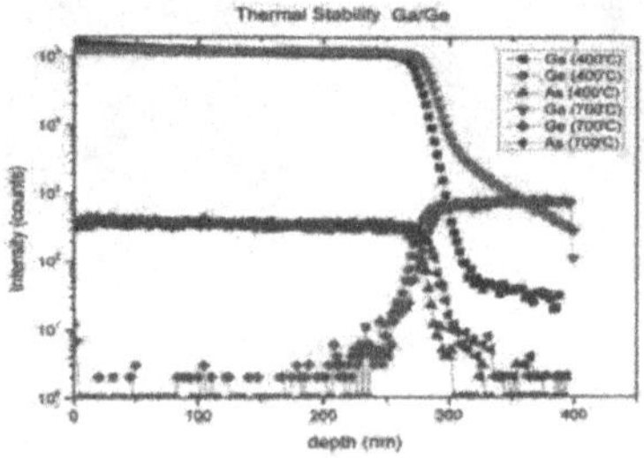

Figure 4. SIMS depth profile of epi-GaAs on Ge after PDA at 400°C and 700°C.

Electrical characterization

The C-V curves at 100kHz for TaN/ZrO$_2$/n-GaAs/Ge MIS capacitors annealed at 300°C, 400°C, 500°C and 600°C are shown in figure 5a together with a control sample with no PDA. The ZrO$_2$ thickness measured from HRTEM image is about 13nm. The MIS capacitor annealed at 400°C shows a clear transition from accumulation to inversion. It has the smallest C-V stretch out and attains the highest accumulation capacitance. Since C-V stretch out is indicative of interfacial defects [11], the sample annealed at 400°C should have the best ZrO$_2$/GaAs interface quality. By contrast, the 500°C and 600°C samples have more stretch out and exhibit a decrease in accumulation capacitance. This suggests a more defective interfacial layer has formed. For the 300°C sample, the stretch out is similar to the 400°C sample but the accumulation capacitance is lower. This shows that for optimum electrical characteristics, a 400°C PDA should be used.

The variation of the inversion capacitance in figure 5a is due to the diffusion of Ge into GaAs as illustrated in figure 4. For different PDA temperature, the extent of Ge autodoping in GaAs will differ and this can cause the high frequency inversion capacitance to vary. There is also a capacitance maximum at ~1.25V in the C-V curve of the as-deposited sample which is not seen in the annealed samples. Since ZrO$_2$ was deposited by sputtering, this anomalous feature should be associated with defects due to energetic ions. Further study, however, is needed to clarify this feature.

In figure 5b, we compare the C-V characteristics at 10kHz of the TaN/ZrO$_2$/n-GaAs/Ge sample annealed at 400°C with similar MIS capacitors fabricated on n-GaAs and p-GaAs bulk wafers and annealed at 400°C. The thickness of the ZrO$_2$ layer is the same for each sample. The hysteresis voltage defined as the voltage shift at the same capacitance during the forward and reverse voltage sweeps is greatest for the TaN/ZrO$_2$/n-GaAs/Ge capacitor (~1.3V). The maximum accumulation capacitance is also smaller than that of the TaN/ZrO$_2$/p-GaAs capacitor. The maximum accumulation capacitance of the TaN/ZrO$_2$/n-GaAs capacitor is similar to the TaN/ZrO$_2$/n-GaAs/Ge capacitor but the hysteresis is slightly smaller. Thus, MIS capacitors

122

formed on bulk GaAs has better high-k/III-V interface quality than the epilayer samples. This is due to the increased surface roughness of the epilayer and the sputtering process.

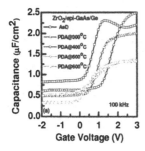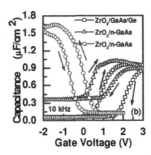

Figure 5. (a) C-V characteristics at 100kHz of TaN/ZrO$_2$/epi-GaAs/Ge MIS capacitors annealed at 300°C, 400°C, 500°C, 600°C by rapid thermal annealing and control sample; (b) C-V characteristics at 10kHz of ZrO$_2$/GaAs/Ge, ZrO$_2$/n-GaAs and ZrO$_2$/p-GaAs MIS capacitors.

Figure 6a shows the frequency dispersion of the C-V curves for the TaN/ZrO$_2$/n-GaAs/Ge capacitor from 10kHz-100kHz. The change in accumulation capacitance at a given voltage is about 50%. This is greater than the dispersion of the TaN/ZrO$_2$/p-GaAs capacitor (25%) (figure 6c). The degradation of the C-V dispersion characteristics in figure 6b should be due to increased leakage in the ZrO$_2$ in this sample because of the sputter deposition. Frequency dispersion occurs because the electrically active defects at the ZrO$_2$/p-GaAs and the ZrO$_2$/epi-GaAs interface have a finite frequency response. As the frequency of the applied field increases, the capacitance associated with these defects will decrease. Since the equivalent circuit of a non-ideal MIS capacitor has the interface trap capacitance in parallel with the channel capacitance, the overall MIS capacitance decreases.

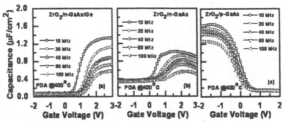

Figure 6. Frequency dispersion (10-100kHz) of C-V characteristics of (a) ZrO$_2$/n-GaAs/Ge, (b) ZrO$_2$/n-GaAs and (c) ZrO$_2$/p-GaAs.

The frequency dispersion of the C-V characteristics of the TaN/HfO$_2$/n-GaAs/Ge capacitor from 10kHz to 100kHz was also measured [12]. Unlike figure 6a, there is a much smaller dispersion of only 10%. Hence, despite the rougher surface morphology of the n-GaAs epilayer, the ALD process is able to form a good quality interface. This is because the ALD process is surface reaction limited and ALD occurs conformally. By contrast, the sputtering process involves energetic ions and their impingement can result in more defects in the interfacial layer.

CONCLUSIONS

We have demonstrated the epitaxial growth of GaAs by MOCVD on Ge substrates with potential III-V n-MOSFET application. The GaAs/Ge interface was found to be of good quality with low defect levels. During epitaxy at 620°C, Ge diffuses up to 60nm into GaAs and during PDA, there is some Ga diffusion into Ge. The C-V characteristics of the sputtered ZrO_2/epi-GaAs interface annealed at 400°C show greater hysteresis, lower accumulation capacitance and greater frequency dispersion than that of ZrO_2/ bulk GaAs interfaces. This is due to the presence of a thicker interfacial layer with defects and can be reduced by using ALD high-k dielectrics.

REFERENCES

1. S.E. Thompson, R.S. Chau, T. Ghani, K. Mistry and S. Tyagi, *IEEE Trans. Semicond. Manu.* **18**, 26 (2005).
2. M. Heyns and W. Tsai, *MRS Bull.* **34**, 485 (2009).
3. Y. Gong, C.M. Ng and T.K.S. Wong, *J. Electrochem. Soc.* **156**, H948 (2009).
4. F. Ren, M. Hong, W.S. Hobson, J.M. Kuo, J.R. Lothian, J.P. Mannaerts, J. Kwo, S.N.G. Chu, Y.K. Chen and A.Y. Cho, *Solid-State Electron.* **41**, 1751 (1997).
5. P.D. Ye, G.D. Wilk and M.M. Frank, "Processing and Characterization of III-V Compound Semiconductor MOSFETs Using Atomic Layer Deposited Gate Dielectrics," *Advanced Gate Stacks for High-Mobility Semiconductors*, ed. A. Dimoulas, E. Gusev, P.C. McIntyre and M. Heyns (Springer, 2007) pp. 341-361.
6. W. Tsai, N. Goel, S. Kovershnikov, P. Majhi and W. Wang, *Microelect. Eng.* **86**, 1540 (2009).
7. Y. Xuan, Y.Q. Wu and P.D. Ye, *IEEE Elect. Dev. Lett.* **29**, 294 (2008).
8. M. Passlack, R. Droopad, K. Rajagopalan, J. Abrokwah, P. Zurcher, R. Hill, D. Moran, X. Li, H. Zhou, D. MacIntyre, S. Thoms, I. Thayne, Dig. CS MANTECH Conf. Austin, TX, 2007 pp. 235-238.
9. S.M. Sze, *Physics of Semiconductor Devices*, (Wiley, 1981) pp. 850-851.
10. C.O. Chui and K.C. Saraswat, "Germanium Nanodevices and Technology," *Advanced Gate Stacks for High-Mobility Semiconductors,* ed. A. Dimoulas, E. Gusev, P.C. McIntyre and M. Heyns (Springer, 2007) pp. 293-313.
11. R. Degraeve, E. Cartier, T. Kauerauf, R. Carter, L. Pantisano, A. Kerber and G. Groseneken, *MRS Bull.* **27**, 222 (2002).
12. G.K. Dalapati, M.K. Kumar, C.K. Chia, H. Gao, B.Z. Wang, A.S.W. Wong, A. Kumar, S.Y. Chiam, J.S. Pan and D.Z. Chi, *J. Electrochem. Soc.* **157**, H825 (2010).

Mater. Res. Soc. Symp. Proc. Vol. 1252 © 2010 Materials Research Society   1252-I07-08

Interface study of SiO$_2$/ HfO$_2$/SiO$_2$ stacks used as InterPoly Dielectric for future generations of embedded Flash memories.

A. Guiraud [1,2] , N. Breil [3], M. Gros-Jean [1], D. Deleruyelle [2] , G. Micolau [2] , C. Muller [2],
N. Chérault [1] and P. Morin [1]
[1] ST Microelectronics, 850 rue Jean Monnet, 38926 Crolles, France
[2] IM2NP UMR CNRS 6242, IMT Technopôle de Château Gombert, 13451 Marseille Cedex 20, France
[3] IBM Microelectronics, 850 rue Jean Monnet, 38926 Crolles, France

ABSTRACT

We have investigated the integration of Hf-based material as Inter Poly Dielectric in flash memories devices. Electrical measurements showed the good properties of SiO$_2$/HfO$_2$/SiO$_2$ stacks. We then interested to the impact of the thermal budget on this specific stack which induces changes in the electrical properties. XPS measurements suggests those changes are due to the presence of an Hf-silicate layer at the SiO$_2$/HfO$_2$ interface.

INTRODUCTION:

The Flash architecture consists in a MOSFET with a floating gate buried in the oxide. The floating gate is insulated from the silicon by the tunnel oxide and insulated from the control gate by the Inter Poly Dielectric (IPD). The Oxide-Nitride-Oxide (ONO) stack is currently used in Flash memory technology as IPD. Its role is to ensure a good coupling ratio between the control gate and the floating gate allowing program/erase trough tunnel oxide. For future generations of Non-Volatiles Memories (NVM), ONO stack almost reached its scaling limit in terms of leakage current and the use of high-k materials is mandatory in order to maintain (or improve) the coupling ratio after down scaling [1]. For embedded NVM (eNVM) applications, the process must be CMOS compatible. Materials like HfO$_2$ and its silicates (HfSiO, HfSiON) used for high-κ metal gate technology seems interesting to use. We keep the top and bottom SiO$_2$ layers because of the high electron barrier for retention in temperature issues as well as a good interface with the poly-Si.

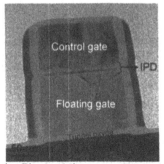

Fig.1: Transmission Electron Microscopy (TEM) view of a flash cell

In this paper, we study electrical of SiO₂/High-k/SiO₂ stacks an try to establish links with physical properties.

EXPERIMENT

The analyzed samples were MIS capacitors composed of a tri-layer $SiO_2/HfO_2/SiO_2$, $SiO_2/HfSiO/SiO_2$ or $SiO_2/HfSiON/SiO_2$ stacks, on the top of which TiN electrodes were patterned. Prior to the high-k material deposition, a 5nm thick High Temperature Oxyde (HTO) SiO_2 layer was deposited at 750°C on p-type (100) silicon wafer. The HfO_2 layer was then grown by Plasma Enhanced Atomic Layer Deposition (PEALD) using Tetrakis[EthylMethylAmino]Hafnium (TEMAH) precursors and O_2 plasma at 250°C. HfSiO layer was deposited by Metal-Organic Chemical Vapor Deposition (MOCVD) at 600°C, Decoupled Plasma Nitridation (DPN) followed by Post Nitridation Anneal (PNA) was performed on some samples to form HfSiON. The high-k layer was subsequently capped with a second 5nm thick HTO SiO_2 layer. References ONO samples were processed by Low Pressure Chemical Vapor Deposition (LPCVD). A Rapid Thermal Anneal (RTA) for 15s at 1100°C under N_2+O_2 (95:5) was performed on some samples after HTO deposition to simulate the higher thermal budget experienced by IPD during the eNVM process. The TiN material used as top electrode was finally deposited by ALD using $TiCl_4$ and NH_3 precursors at 400°C. The electrodes, patterned by dry etching, have a broad range of surfaces from $100 \times 100 \mu m^2$ to $1000 \times 600 \mu m^2$. J-V measurements were performed with parameter analyzer (HP4156C) with 0 to -20V voltage ramp, -100mV step. C-V were performed with LCR-meter (HP4284A) at 10kHz from -5 to 2V with 25mV alternative signal. Measures were taken with contacts on the top electrode and on the bulk and were performed on several electrodes to be sure of the reproducibility. XPS were performed with Al source Kα 1486.6 eV.

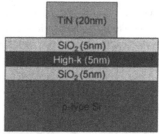

Fig 2 : Sample description : a SiO₂ / High-k/ SiO₂ is deposited on a bare silicon, potentially annealed. Then, the TiN is deposited and patterned.

RESULTS

Electrical results

We have made electrical measurements on SiO₂/High-κ/SiO₂ metal gate stack capacitors, the high-κ layer being composed of HfO_2, HfSiO or HfSiON which are already usual materials for the most advanced CMOS applications. ONO capacitors were also tested as a reference. In order to simulate the impact of thermal budgets related to the subsequent logic transistors processing, a rapid thermal annealing at 1100°C during 15sec was performed on all samples. Both the bottom oxide and top oxide have same thicknesses for all samples. The High-K layer thickness was chosen in order to obtain a comparable EOT for all samples, which assumption is fulfilled as demonstrated by the C-V measurements shown in Fig. 3.

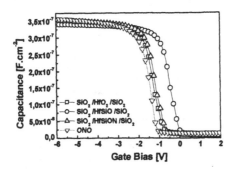

Fig.3 : Capacitance measurement of SiO_2/Hf-based/SiO_2 and ONO stacks

J-V measurements of the different annealed samples are presented in Fig. 4,. It demonstrates that all Hf-based high-k samples have a lower leakage current than ONO. It is interesting to note that the HfO_2 shows the lowest leakage current at either low or high field, the both of them being of paramount importance for flash cell performance. As a consequence, the following results are mainly focused on the $SiO_2/HfO_2/SiO_2$ stack.

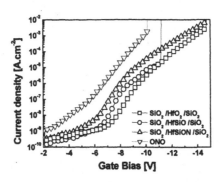

Fig.4 : Leakage measurement of SiO_2/Hf-based/SiO_2 and ONO stacks

In Fig. 5 we focus our investigations on the impact of the high temperature anneal on the $SiO_2/HfO_2/SiO_2$ stack. Both as deposited and annealed stacks have roughly the same EOT value. According to the literature, HfO_2 crystallization occurs at 550°C [2], and in our case will be induced by the top oxide deposition performed at 750°C. The subsequent high temperature annealing has no impact on the EOT.

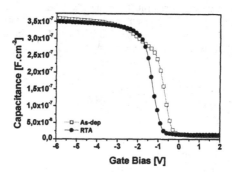

Fig.5 : Capacitance measurement of $SiO_2/HfO_2/SiO_2$ stacks with and without RTA

When comparing those data with the EOT calculated from the different layers dielectric constants and thicknesses, it appears that the measured EOT is higher than expected. Also, we observe a Vth shift toward negative bias indicating an increase of positive fixed charges. Those changes suggest a modification of the HfO_2 bulk structure, and/or some reactions at the SiO_2/HfO_2 interfaces.

Physical characterization results

In order to have a XPS depth profile for interface investigation, we performed a wet etching of the top oxide using a diluted (2.5%) fluoridric acid (HF) chemistry. We noticed an immunity of the HfO_2 to the HF etching, which is the sign of High-k layer crystallization. This phenomenon has been observed by Besson et al. [7].

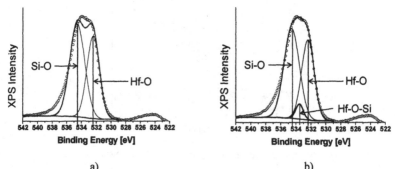

a)   b)

Fig. 6: O1s XPS spectra, fitting the peak at interface with only O-Si and O-Hf bonds (a), and with O-Si, O-Hf and Hf-O-Si bonds (b).

XPS measurements were performed on the annealed stack with the top oxide partially removed, and are presented on Fig. 6. The O1s spectra can not be associated only to a convolution of O-Si and O-Hf peaks (Fig. 6a). Fitting is clearly improved when adding a third peak corresponding to the Hf-O-Si bond (Fig. 6b) [3].
This observation suggests the formation of a HfSiO compound at the SiO_2/HfO_2 interface.

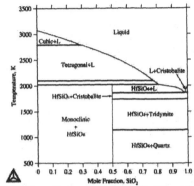

Fig.7 : Phase diagram of SiO_2 and HfO_2 (Shin et al.)

This hypothesis of an interfacial layer formation can be supported by thermodynamic calculations found in literature [4], demonstrating the instability of the SiO_2-HfO_2 system (Fig.7), and the subsequent formation of $HfSiO_4$. Some complementary investigations are on-going in order to quantify the formation of the silicate.

DISCUSSION

Based on the electrical measurements on $SiO_2/HfO_2/SiO_2$ stacks with and without RTA, we observed an impact of the anneal. The increase of the positive charges noticed in the C-V measurements aren't likely due to bulk modifications because the anneal generally passivates the oxygen vacancies. It has been reported [5] that interfaces are rich in defects, thus increasing the positives fixed charges. Those changes were investigated via physical analysis assuming the RTA process induced bulk modifications or interface reactions.
The immunity of the HfO_2 layer to the HF chemistry after anneal is an indication of a monoclinic phase and so a lower k of the HfO_2 (~15). However, according to TEM observations and C-V measurements, the calculated EOT is still inferior to the measured one. Based on this statement, the hypothesis of bulk modification cannot be enough to explain those changes and we explored the hypothesis of interface reactions.
We performed XPS on $SiO_2/HfO_2/Si$ stacks with full HF removal of the SiO_2, partial removal and no etch, in order to have a range of top SiO_2 thicknesses as well as O-Si and O-Hf bonds references. We observed that the O1s spectra peak on partially etched sample is between the O-Si and O-Hf spectras but isn't a convolution of both. Indeed, we can't fit this peak with only O-Si and O-Hf bonds and need to introduce a third peak which relative position regarding HfO_2 and SiO_2 correspond to Hf-O-Si [3]. Adding this peak improve the fitting and is an other clue of

probable interface reaction. Thermodynamics calculations found in literature [4] indicates there is no stable phase between SiO_2 and HfO_2 and a formation of $HfSiO_4$ interlayer. It is reported that interfacial silicate layer is Si-rich and has a low dielectric constant [6], which may explain the increase of the EOT.

CONCLUSION:

We suggest a possible reaction between SiO_2 an HfO_2 under high temperature anneal that can cause modifications of the electrical behavior of the $SiO_2/HfO_2/SiO_2$ stack which appears to be a promising candidate for IPD applications.

REFERENCE:

[1] D. Wellekens, J.V. Houdt, and S. Member, "The Future of Flash Memory : is Floating Gate Technology Doomed to Lose The Race ?," 2008.
[2] M. Modreanu, J. Sancho-parramon, D.O. Connell, and J. Justice, "Solid phase crystallisation of HfO2 thin films," *Materials Science and Engineering B*, vol. 118, 2005, pp. 127-131.
[3] G. He, L.D. Zhang, and Q. Fang, "Silicate layer formation at HfO_2 / SiO_2 / Si interface determined by x-ray photoelectron spectroscopy and infrared spectroscopy," *Journal of Applied Physics*, vol. 100, 2006, pp. 1-5.
[4] D. Shin, R. Arroyave, and Z. Liu, "Thermodynamic modeling of the Hf − Si − O system" *Computer Coupling of phase diagramm and thermochemistry*, vol. 30, 2006, pp. 375-386.
[5] Z. Zhang, M. Li, and S.A. Campbell, "Effects of Annealing on Charge in HfO 2 Gate Stacks", *IEEE Electron Device Letter*, vol. 26, 2005, pp. 20-22.
[6] N. Barrett, O. Renault, J. Damlencourt, and F. Martin, "metal-oxide-semiconductor gate oxide stacks : A valence band and quantitative core-level study by soft x-ray photoelectron spectroscopy," *Journal of Applied Physics*, vol. 96, 2005, pp. 6362-6369.
[7] P. Besson, V. Loup, T. Salvetat, N. Rochat, S. Lhostis, S. Favier, K. Dabertrand and V. Cosnier, "Critical thickness threshold in HfO_2 layers", *Solid State Phenomena*, vol. 134, pp. 67-70.

Materials and Devices for Beyond CMOS Scaling

Mater. Res. Soc. Symp. Proc. Vol. 1252 © 2010 Materials Research Society   1252-J02-08

Fabrication and Current-Voltage Characteristics of Ni Spin Quantum Cross Devices with P3HT:PCBM Organic Materials

Hideo Kaiju[1, 2], Kenji Kondo[1], Nubla Basheer[1], Nobuyoshi Kawaguchi[1], Susanne White[1], Akihiko Hirata[3], Manabu Ishimaru[3], Yoshihiko Hirotsu[3], and Akira Ishibashi[1]
[1] Laboratory of Quantum Electronics, Research Institute for Electronic Science, Hokkaido University, Sapporo 001-0020, Japan
[2] PRESTO, Japan Science and Technology Agency, Saitama 332-0012, Japan
[3] The Institute of Scientific and Industrial Research, Osaka University, Ibaraki, Osaka 567-0047, Japan

ABSTRACT

We have proposed spin quantum cross (SQC) devices, in which organic materials are sandwiched between two edges of magnetic thin films whose edges are crossed, towards the realization of novel beyond-CMOS switching devices. In SQC devices, nanometer-size junctions can be produced since the junction area is determined by the film thickness. In this study, we have fabricated Ni SQC devices with poly-3-hexylthiophene (P3HT): 6, 6-phenyl C61-butyric acid methyl ester (PCBM) organic materials and investigated the current-voltage (I-V) characteristics experimentally and theoretically. As a result of I-V measurements, ohmic I-V characteristics have been obtained at room temperature for Ni SQC devices with P3HT:PCBM organic materials, where the junction area is as small as 16 nm x 16 nm. This experimental result shows quantitative agreement with the theoretical calculation results performed within the framework of the Anderson model under the strong coupling limit. Our calculation also shows that a high on/off ratio beyond 10000:1 can be obtained in Ni SQC devices with P3HT:PCBM organic materials under the weak coupling condition.

INTRODUCTION

Molecular electronics have stimulated considerable interest as a technology that may enable a next generation of high-density memory devices [1,2]. Especially, in International Technology Roadmap for Semiconductor (ITRS) 2009 edition, molecular memory devices have been expected as candidates for beyond-CMOS devices since they offer the possibility of nanometer-scale components [3]. Recently, we have proposed spin quantum cross (SQC) devices, in which organic materials are sandwiched between two edges of magnetic thin films whose edges are crossed, towards the realization of novel beyond-CMOS switching devices [4-7]. In SQC devices, the area of the crossed section is determined by the film thickness, in other words 1-20 nm thick films could produce 1×1-20×20 nm^2 nanoscale junctions. This method offers a way to overcome the feature size limit of conventional optical lithography and to realize switching devices with a high on/off ratio. Moreover, the resistance of the electrode can be reduced down to ~kΩ since the width of films can be easily controlled to the one as long as ~mm. This makes it possible to detect the resistance of the junction with high sensitivity and to be applied to high-frequency devices. Thus, SQC devices with organic materials can be expected as novel beyond-CMOS switching devices with high on/off ratios and low-resistance electrodes. In this study, towards the creation of such novel beyond-CMOS devices, we have investigated the resistance properties of Ni electrodes on polyethylene naphthalate (PEN) substrates used in SQC devices and the current-voltage (I-V) characteristics of Ni SQC devices with poly-3-hexylthiophene (P3HT): 6, 6-phenyl C61-butyric acid methyl ester (PCBM) organic materials experimentally and theoretically.

EXPERIMENTS

The fabrication method of SQC devices is shown in figure 1. First, Ni thin films have been thermally evaporated on PEN substrates (2 mm width, 10 mm length, and 20 μm thickness) in a high vacuum chamber at a base pressure of ~10^{-8} torr. The pressure during the evaporation is 10^{-5} torr and the temperature near PEN substrates is less than 62 °C, which is lower than the glass transition temperature T_g of 120 °C for PEN substrates. The growth rate is 0.93 nm/min at an evaporation power of 350 W. Then, fabricated Ni/PEN films have been sandwiched between two polymethyl methacrylate (PMMA) resins and the edge of the PMMA/Ni/PEN/PMMA structure has been polished by chemical mechanical polishing (CMP) methods using alumina (Al_2O_3) slurries with particle diameters of 0.1, 0.3, and 1.0 μm. Finally, P3HT:PCBM organic materials have been sandwiched between two sets of PMMA/Ni/PEN/PMMA structures whose edges are crossed. The Ni thickness has been measured by an optical method using the diode pumped solid state green laser and the photo diode detector. The microstructures as well as the Ni/PEN interfacial structures have been examined using transmission electron microscopy (TEM) and electron diffraction (ED). The cross-sectional TEM samples have been prepared by a combination of mechanical polishing and Ar ion thinning. The resistance of Ni electrodes has been measured by a two-probe method at room temperature. The I-V characteristics of Ni SQC devices with P3HT:PCBM organic materials have been measured by a four-probe method at room temperature.

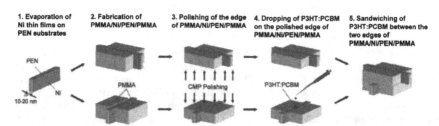

Figure 1. Fabrication method of Ni SQC devices with P3HT:PCBM organic materials.

RESULTS AND DISCUSSION

Figure 2(a) shows the Ni thickness dependence of the electric resistivity for Ni thin films on PEN substrates. The electric resistivity ρ_{Ni} increases with decreasing the Ni thickness. In order to

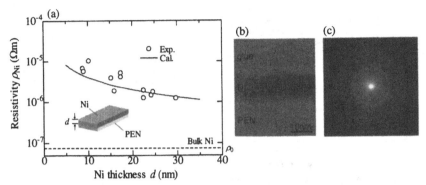

Figure 2. (a) Ni thickness dependence of the electric resistivity for Ni/PEN. (b) High-resolution TEM image and (c) ED pattern for Ni (11nm)/PEN.

explain this experimental result quantitatively, we have calculated the electric resistivity using Mayadas-Shatzkes model [8]. According to Mayadas-Shatzkes model, the electric resistivity ρ_{Ni} is expressed by

$$\rho_{Ni} / \rho_0 = \left[1 - \frac{3}{2}\alpha + 3\alpha^2 - 3\alpha^3 \ln(1 + \frac{1}{\alpha})\right]^{-1}, \qquad (1)$$

$$\alpha = \frac{\lambda}{D}\frac{R}{1-R}, \qquad (2)$$

where λ is the electron mean free path, D is the average grain diameter, R is the reflection coefficient for electrons striking the grain boundary, and ρ_0 is the electric resistivity for bulk Ni. The electron mean free path λ is 11 nm for bulk Ni. The average grain diameter D is 3 nm, which has been obtained from high-resolution TEM image and ED pattern shown in figure 2(b) and (c). The reflection coefficient R is 0.71-0.95, which is the extrapolation value obtained from R in Ni thin films with the thickness of 31-115 nm [9]. From figure 2(a), the experimental result shows good agreement with the calculation result quantitatively. This means that the main contribution to the electric resistivity comes from the electron scattering at grain boundaries in Ni thin films on PEN substrates. Here, we discuss the feasibility of Ni thin films on PEN substrates for the electrodes of SQC devices. As can be seen from figure 2(a), the electric resistivity of Ni thin films on PEN substrates is 1-2 orders larger than that of bulk Ni. However, as we mentioned in the introduction section, the electrode resistance can be reduced since the film width can be controlled to the one as long as ~mm. According to a simple estimation, low-resistance electrodes with 0.1-1 kΩ can be obtained when 10-30 nm thick films are used. Based on this estimation, we have fabricated Ni SQC devices with P3HT:PCBM organic materials and investigated the resistance of Ni electrodes. Figure 3(a) shows the Ni electrode resistance as a function of the line width l, which corresponds to the Ni thickness d in SQC devices. The schematic illustration of SQC devices is shown in figure 3(b). In figure 3(a), the Ni electrode resistances in the conventional cross-bar structures are also shown. The black solid line, broken

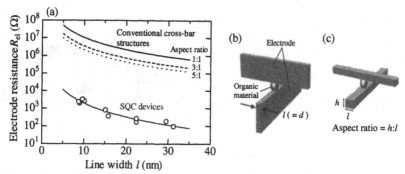

Figure 3. (a) Electrode resistance of SQC devices and conventional cross-bar structures. Schematic illustration of (b) SQC devices and (c) conventional cross-bar structures.

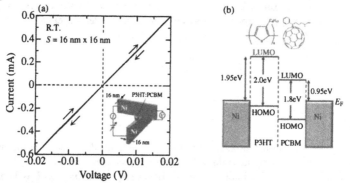

Figure 4. (a) *I-V* characteristics and (b) energy diagram for Ni SQCS devices with P3HT: PCBM organic materials.

line, and dotted line represent the Ni electrode resistance estimated in the conventional cross-bar structures with an aspect ratio of 1:1, 3:1, and 5:1, respectively. The schematic illustration of conventional cross-bar structures is shown in figure 3(c). From figure 3(a), the Ni electrode resistance in the conventional cross-bar structures is as large as 1-10 MΩ. In contrast, it is found that the low-resistance electrode with 0.1-1 kΩ can be realized in SQC devices.

Then, we have measured the *I-V* characteristics of SQC devices using these low-resistance electrodes. Figure 4(a) shows the *I-V* characteristics for Ni SQC devices with P3HT:PCBM organic materials at room temperature. The Ni thickness is 16 nm. Therefore, the junction area is as small as 16 nm x 16 nm. We have obtained ohmic characteristics with a junction resistance of 32 Ω. In order to explain this experimental result, we have calculated the *I-V* characteristics of SQC devices within the framework of the Anderson model. The current flowing across the junction in SQC devices can be expressed by

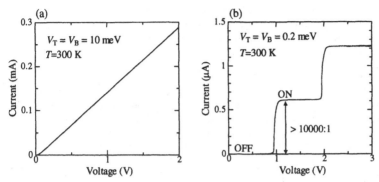

Figure 5. Calculated *I-V* characteristics of Ni SQC devices with P3HT:PCBM organic materials (a) under the strong coupling limit and (b) under the weak coupling condition.

$$I = \frac{2e^2}{h} \int_{E_F}^{E_F + eV} d\varepsilon \sum_i \left(\frac{4\Gamma_T(\varepsilon)\Gamma_B(\varepsilon)}{(\varepsilon - \varepsilon_0(i))^2 + (\Gamma_T(\varepsilon) + \Gamma_B(\varepsilon))^2} \right) [f(\varepsilon - eV - E_F) \cdot f(\varepsilon - E_F)] , \qquad (3)$$

where e is the elementary charge, h is the Planck's constant, E_F is the Fermi energy of Ni, $f(\varepsilon)$ is the Fermi-Dirac distribution function, and $\varepsilon_0(i)$ is the i-th energy level of eigen-states for the P3HT:PCBM organic material [7]. We used the value of 9.071 eV for the Ni Fermi level [10], and also assumed that the P3HT:PCBM organic material had two energy levels of $\varepsilon_0(1) = 0.95$ eV and $\varepsilon_0(2) = 1.95$ eV, estimated from the Ni Fermi level, as shown in figure 4(b)[11,12]. $\Gamma_{T(B)}$ is the coupling strength between the Ni top (bottom) electrode and the P3HT:PCBM organic material, which is given by

$$\Gamma_{T(B)}(\varepsilon) = \pi D_{T(B)}(\varepsilon) |V_{T(B)}|^2 , \qquad (4)$$

where $D_{T(B)}$ is a density of states of electrons for the Ni top (bottom) electrode and $V_{T(B)}$ is the coupling constant between the Ni top (bottom) electrode and the P3HT:PCBM organic material. Figure 5(a) shows the calculated *I-V* characteristics of SQC devices under the strong coupling limit. $V_{T(B)}$ is assumed to be 10.0 meV, corresponding to $\Gamma_{T(B)}$ of 3927 meV. We have obtained the ohmic *I-V* characteristics with a resistance of 6.7 kΩ. In this calculation, the junction area is 1 nm x 1 nm, which is expected as a size of one P3HT:PCBM organic molecule, and the number of the conductance channel is four, taking into consideration the spin degeneracy. In experiments, the junction area of P3HT:PCBM organic materials is 16 nm x 16 nm, which corresponds to 1024 (=4x16x16) conductance channels. Therefore, the junction resistance in a size of 16 nm x 16 nm is calculated to be 26 Ω (=6.7kΩ/16/16), which is in good agreement with the experimental value of 32 Ω.

Finally, we have calculated the *I-V* characteristics of SQC devices under the weak coupling condition, shown in figure 5(b). $V_{T(B)}$ is assumed to be 0.2 meV, corresponding to $\Gamma_{T(B)}$ of 1.57 meV. From figure 5(b), the calculated result shows the sharp steps at the positions of the energy level of the P3HT:PCBM organic material. The off-state current I_0 is 34.1 pA at the voltage V_0

of 0.1 V, and the on-state current I_1 is 0.59 μA at the voltage V_1 of 1.05 V. As we estimate the switching on/off ratio, the I_1/I_0 ratio is found to be an excess of 10000:1. This indicates that SQC devices under the weak coupling condition can be expected to have potential application in novel switching devices with a high switching ratio.

CONCLUSIONS

We have fabricated Ni SQC devices with P3HT:PCBM organic materials and investigated the I-V characteristics experimentally and theoretically. As a result of I-V measurements, ohmic I-V characteristics have been obtained at room temperature for Ni SQC devices with P3HT:PCBM organic materials, where the junction area is as small as 16 nm x 16 nm. This experimental result shows quantitative agreement with the theoretical calculation results performed within the framework of the Anderson model under the strong coupling limit. Our calculation also demonstrates that a high on/off ratio beyond 10000:1 can be obtained in Ni SQC devices with P3HT:PCBM organic materials under the weak coupling condition.

ACKNOWLEDGMENTS

This research has been partially supported by Special Education and Research Expenses from Post-Silicon Materials and Devices Research Alliance, a Grant-in-Aid for Young Scientists from MEXT, Precursory Research for Embryonic Science and Technology program and Research for Promoting Technological Seeds from JST, Foundation Advanced Technology Institute (ATI), and a Grant-in-Aid for Scientific Research from JSPS. The authors would like to express their sincere appreciation to Dr. M. Hirasaka of Teijin Ltd., Research Manager K. Kubo of Teijin DuPont Films Japan Ltd., Prof. M. Yamamoto, Assist. Prof. K. Matsuda, Dr. S. Jin, H. Sato and M. Takei in Hokkaido University for helpful discussions.

REFERENCES

1. J. Chen, M. A. Reed, A. M. Rawlett, and J. M. Tour, Science **286**, 1550(1999).
2. Y. Chen, D. A. A. Ohlberg, X. Li, D. R. Stewart, R. S. Williams, J. O. Jeppesen, K. A. Nielsen, J. F. Stoddard, D. L. Olynick, and E. Anderson, Appl. Phys. Lett. **82**, 1610 (2003).
3. Semiconductor Industry Association, International Technology Roadmap for Semiconductors (ITRS) 2009 ed.
4. K. Kondo and A. Ishibashi, Jpn. J. Appl. Phys. **45**, 9137 (2006).
5. H. Kaiju, A. Ono, N. Kawaguchi, and A. Ishibashi: J. Appl. Phys. **103**, 07B523 (2008).
6. H. Kaiju, A. Ono, N. Kawaguchi, K. Kondo, A. Ishibashi, J. H. Won, A. Hirata, M. Ishimaru, and Y. Hirotsu: Appl. Sur. Sci. **255**, 3706 (2009).
7. K. Kondo, H. Kaiju, and A. Ishibashi, J. Appl. Phys. **105**, 07D5221 (2009).
8. A. F. Mayadas and M. Shatzkes: Phys. Rev. B **1**, 1382 (1970).
9. C. Nacereddine, A. Layadi, A. Guittoum, S. −M. Cherif, T. Chauveau, D. Billet, J. B. Youssef, A. Bourzami, and M. −H. Bourahli: Mater. Sci. Eng. B **136**, 197 (2007).
10. C. S. Wang and J. Callaway, Phys. Rev. B **15**, 298 (1977).
11. D. E. Eastman, Phys. Rev. B **2**, 1 (1970).
12. B. C. Thompson and J. M. J. Frecht, Angew. Chem. Int. Ed. **47**, 58 (2008).

Mater. Res. Soc. Symp. Proc. Vol. 1252 © 2010 Materials Research Society   1252-J03-01

Optically Active Defects in an InAsP/InP Quantum Well Monolithically Integrated on SrTiO₃ (001)

J. Cheng[1], A. El Akra[1], C. Bru-Chevallier[1], G. Patriarche[2], L. Largeau[2], P. Regreny[1], G. Hollinger[1] and G. Saint-Girons[1]

[1] Université de Lyon; Institut des Nanotechnologies de Lyon INL-UMR5270-CNRS, Ecole Centrale de Lyon, 36 avenue Guy de Collongue, 69134 Ecully, France
[2] LPN-UPR20/CNRS, route de Nozay, 91460 Marcoussis, France

ABSTRACT

The optical properties of an InAsP/InP quantum well grown on a SrTiO₃(001) substrate are analyzed. At 13K, the photoluminescence (PL) yield of the well is comparable to that of a reference well grown on an InP substrate. Increasing the temperature leads to the activation of non-radiative mechanisms for the sample grown on SrTiO₃. The main non-radiative channel is related to the thermal excitation of the holes in the first heavy hole excited state, followed by the non-radiative recombination of charge carriers on twins and/or domain boundaries, in the immediate vicinity of the well.

INTRODUCTION

Integrating III-V semiconductors on silicon would allow not only combining optoelectronic functionalities with standard Si-based CMOS systems, but also envisaging the use of high mobility III-V channels for fabricating high speed P and N-MOS transistors[1]. However, the direct growth of III-V on Si is always limited by a too large lattice mismatch. Overcoming this limitation has motivated numerous researches in the past 20 years[2,3,4]. Various strategies of heterogeneous integration have been proposed, mostly based on wafer fusion of molecular bonding techniques, possibly associated to Smartcut processes[5]. These techniques have reached an advanced degree of technological maturity, but are still limited by their technological complexity and cost.

In recent years, the monolithic integration of crystalline SrTiO₃ (STO) on silicon has motivated numerous researches.[6,7] In the early 2000's, Motolora has published interesting but controversial results concerning the monolithic integration of GaAs based transistors on Si using crystalline SrTiO₃/Si templates[8]. On this basis, several groups have proposed using Pr₂O₃/Si(111)[9], or SrHfO₃/Si(001)[10] templates for the monolithic integration of Ge on Si. In our recent studies, we have described the peculiar behavior of III-V/oxide interface: the III-V material nucleates with its bulk lattice parameter as soon as the growth begins and the mismatch is fully accommodated by a regular array of dislocations confined at the heterointerface[11]. As a consequence, the growing III-V material does not contain threading dislocations due to plastic relaxation. We have used this peculiarity to grow InP based heterostructures on Gd₂O₃/Si (111)[12] templates and STO (001) substrates[11]. We have also identified the main challenges related to the

139

growth of InP on STO, namely a relatively large interface energy and the formation of specific defects such as microtwins and antiphase boundaries[13].

In this letter, we propose a study of the influence of these defects on the optical properties of an InAsP/InP quantum well grown on a STO(001) substrate. Temperature-dependent photoluminescence (PL) experiments combined with transmission electron microscopy (TEM) and atomic force microscopy (AFM) measurements allow identifying the main non-radiative channels in the quantum well.

EXPERIMENT

The sample considered in the following was grown on a STO(001) substrate by solid-source molecular beam epitaxy (MBE). The substrate was first etched in a buffered HF solution and then annealed at 600°C under ultra-vacuum during 30 minutes, leading to the formation of a clean (2x1)-reconstructed TiO_2-terminated surface. A 1μm thick InP layer was grown on this surface, at 480°C with a growth rate of 1μm/h and under a phosphorus partial pressure of 10^{-5} Torr. An InAsP quantum well was then grown on this InP layer at 510°C under As and P partial pressures of $4x10^{-6}$ Torr and $2.7x10^{-6}$ Torr, respectively. Afterwards, a 100 nm InP capping layer was grown on the quantum well. According to X-ray diffraction measurements (not shown here), the InAsP quantum well presents a thickness of 7nm and an As composition of 0.52. A similar structure (containing a 6 nm thick $InAs_{0.63}P_{0.37}$ quantum well) was grown on a bulk InP (001) substrate for comparison (reference sample). Both samples were studied by temperature-dependent PL. PL was excited with the 514.5 nm line of an Ar^+ laser, using a 30 mW incident power focused onto a 200 μm wide spot. The sample was mounted in a closed-cycle temperature-controlled He cryostat. The PL signal was dispersed by a monochromator and collected by a cooled InGaAs array detector.

DISCUSSION

The room-temperature and 13 K PL spectra of both samples are compared in Fig.1. At room temperature, the PL emission of the quantum well grown on STO is centered at 1350 nm (0.92 eV). The associated PL intensity is approximately 100 times less than that of the quantum well grown on InP (reference sample, centered at 1460 nm (0.85 eV)). Moreover, the full width at half maximum (FWHM) of the STO quantum well peak is 53 nm (36 meV), which is sensibly larger than that of the reference sample (47 nm (28 meV)). This indicates that structural defects affect the PL emission of the $SrTiO_3$ sample. At 13 K, the PL yield of the STO quantum well is 15 times smaller than that of the reference sample. Moreover, the reference sample exhibits a narrow PL peak, indicating a good optical quality. Oppositely, the PL peak of the STO sample can be deconvoluted into two Gaussian peaks. The presence of a long wavelength shoulder peak in the spectrum is attributed to an anomal accumulation of As in specific zones around the quantum well, leading to local composition inhomogeneities and to a widening of the PL peak. This shoulder peak is no longer detectable in the room-temperature PL spectrum of the $SrTiO_3$ sample, due to thermally activated depopulation of the corresponding electronic states.

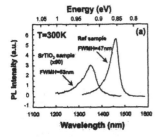 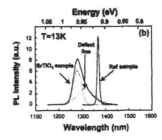

Figure 1: Room temperature (a) and 13 K (b) PL spectra of the reference and STO samples. At room temperature, the PL yield of the STO sample is approximately 100 times less than that of the quantum well of the reference sample

The evolution of the integrated intensity of the quantum well emission as a function of temperature has been measured and normalized to their values at 13K for both samples. The results are displayed in Fig.2(a). Both curves in this figure exhibit the same behavior : the PL intensity remains approximately constant in the low temperature range, and starts decreasing above a given temperature to reach its minimum value at room temperature.

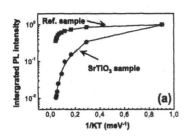 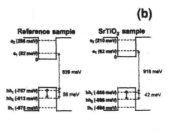

Figure 2: (a) : Evolution of the integrated PL intensity of the STO (circles) and reference (squares) samples as a function of 1/kT (where k is the Boltzmann constant : Arrhenius plot). Continuous lines : curves calculated using the parameters gathered in Table.1. (b) : Calculated band diagrams for the quantum wells in the SrTiO₃ and reference samples.

This behavior is typical for the presence of thermally activated non-radiative defects in the samples. Such evolutions can be fitted using a rate equation based model describing each non-radiative channel by an activation energy and a pre-factor. The activation energy corresponds to the energy to be provided to the charge carriers for them to be thermally excited out of the electronic state corresponding to non-radiative center, and the pre-factor corresponds to the ratio between non-radiative recombination probability in the defect and radiative recombination probability in the quantum well [14,15]. Two non-radiative channels are required to fit the

141

experimental data displayed in Fig.2(a). Thus, the evolution of the PL intensity as a function of temperature obeys following law:

$$I^i(T) = \frac{I^i(0K)}{1 + A_1^i \times \exp\left(-\frac{E_1^i}{k \times T}\right) + A_2^i \times \exp\left(-\frac{E_2^i}{k \times T}\right)},$$ (1)

where the subscripts 1 and 2 designate the first and second non-radiative channels, respectively, and the superscript i (i = ref or i = STO) refers to the sample (reference sample or STO sample). A and E are the above-defined pre-factor and the activation energy, respectively, T is the temperature and k is the Boltzmann constant. The best fits (solid lines in Fig.2(a)) were obtained using the parameters gathered in Table 1.

	A_1	E_1 (meV)	A_2	E_2 (meV)
SrTiO$_3$ sample	19.9	7.9	376.6	42
Reference sample	0.9	5.6	11.1	56

Table 1 : Best fit parameters for the calculated curves of Fig.2(a).

For both samples, the first non-radiative channel is characterized by low activation energies ($E_1^{SrTiO_3} = 7.9$ meV and $E_1^{ref} = 5.6$ meV). These values are very close to that reported for the thickness-dependent exciton binding energies in InGaAs/InP quantum wells. [16] Thus, the first non-radiative channel can be attributed to the dissociation of the excitons trapped in the wells, followed by the non-radiative recombination of the charge carriers in defects located in the wells or at their immediate vicinity, similarly to the mechanism described in Ref.[17]. These defects are more active in the STO sample, as attested by the larger value of $A_1^{SrTiO_3}$ as compared to A_1^{Ref}. The activation energies associated to the second non-radiative channel are larger for both samples ($E_2^{SrTiO_3} = 42$ meV and $E_2^{ref} = 56$ meV). In order to identify the origin of these non-radiative recombinations, the band diagrams of both quantum wells at 300K were calculated (Fig.2(b)) using a 2 bands k.p model, where the effect of the strain in the wells is taken into account by coupling the conduction band and light-hole band [18]. A good match between experimental and calculated emission energies is obtained. Moreover, for both STO and reference samples, the calculated hh$_1$-hh$_2$ transition energies between the heavy hole ground and the first excited states (42 and 56 meV respectively) perfectly match the activation energies associated to the second non-radiative channel. This suggests the following non-radiative mechanism : at low temperature, the holes are confined in the heavy hole ground state. Their probability of presence in the hh$_2$ level increases when increasing the temperature. In this energy level, the extension of their wavefunction out of the well is much larger than in the hh$_1$ level, so that the coupling probability with non-radiative defects located in the immediate vicinity of the well is increased. This effect is at the origin of the second non-radiative channel for both samples. $A_2^{SrTiO_3}$ is much larger than A_2^{ref}, due to the more efficient non-radiative recombinations in the STO sample. These non-radiative recombinations are related to the presence of defects in the STO sample. In a recent study, we have shown that microtwins are formed in InP layers grown on STO(ref 13). The formation of these twins is related to an initially three-dimensional InP growth in the Vollmer–Weber mode. The initially formed InP islands present elevated contact angles[19], and two-dimensional InP layers are formed by lateral growth and coalescence of these islands[20]. The free surfaces of the InP islands are very close to {111} planes, which are known as efficient twinning planes in the Zincblende structure. This enhances twin formation during the lateral growth of the InP islands. Fig.3(a) shows a TEM cross-sectional view of the STO sample in the quantum well region. Microtwins clearly thread the quantum well. Moreover, due to the indirect epitaxial relationship between InP and STO[21], InP is two-domain on STO leading to the presence of antiphase boundaries. At the surface of the STO sample, the width of these antiphase domains reaches approximately 1 μm, as attested by the AFM image of Fig.3(b). As a consequence, domain boundaries thread the quantum well of the STO sample. Twins and domain boundaries are at the origin of the reduction of the PL yield in the STO sample. Holes excited in the hh_2 level of the well are coupled to these non-radiative defects, leading to the above-described PL quenching mechanism. Moreover, the lattice perturbation and roughness related to the presence of these defects locally modify the quantum well composition by creating preferential incorporation sites for arsenic, leading to the observation of a long wavelength peak shoulder in the PL spectrum of the STO sample.

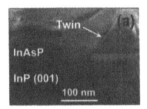 

Figure 3: (a) TEM cross-sectional view of the quantum well of the STO sample. (b) AFM image of the surface of the STO sample. A domain structure can clearly be distinguished on this image.

CONCLUSIONS

Very recently, the same heterostructure as that discussed above has been successfully grown on a STO/Si(001) template. This heterostructure presents similar optical properties as that of the STO sample discussed here. This illustrates the potentialities of our approach for the monolithic integration of InP on Si. However, further optimization is required to improve the PL yield of InP heterostructures grown on Si. It is likely that twin density can be efficiently reduced by finely tuning the InP growth conditions in order to enhance wetting (an ideally obtain a 2D growth mode) at the initial stages of the growth. Avoiding the formation of domain boundaries is more complex, because their origin is related to the geometry of the STO surface. Our further efforts will focus on finding growth conditions leading to a maximization of the domain size.

ACKNOWLEDGMENTS

The authors thank Claude Botella and Jean-Baptiste Goure for technical assistance. This work is supported by the French "Agence Nationale de la Recherche" (BOTOX project #05-JCJC0055, and COMPHETI project ANR-09-NANO-013-01) and by the region Rhone-Alpes micronano cluster (IMOX project)

REFERENCES

[1] P. Singer, Solid State Technology, IEDM (2009)

[2] D. G. Deppe, N. Holonyak, Jr., D. W. Nam, K. C. Hsieh, G. S. Jackson, R. J. Matyi, H. Shichijo, J. E. Epler, H. F. Chung, Appl. Phys. Lett. **51**, 637 (1987)

[3] S.M. Ting and E.A. Fitzgerald, J. Appl. Phys. **87**, 2618, (2000).

[4] M. Kostrzewa, G. Grenet, P. Regreny, J.L. Leclercq, P. Perreau, E. Jalaguier, L. DiCioccio and G. Hollinger, J. Cryst. Growth **275**, 157, (2005).

[5] S. J. B. Yoo, R. Bhat, C. Caneau, and M. A. Koza, Appl. Phys. Lett. **66**, 3410, (1995)

[6] Y. Wang, C. Ganpule, B. T. Liu, H. Li, K. Mori, B. Hill, M. Wuttig, R. Ramesh, J. Finder, Z. Yu, R. Droopad, and K. Eisenbeiser, Appl. Phys. Lett. **80**, 97, (2002)

[7] G. J. Norga, C. Marchiori, C. Rossel, A. Guiller, J. Appl. Phys. **99**, 084102 (2006).

[8] K. Eisenbeiser, R. Emrick, R. Droopad, Z. Yu, J. Finder, S. Rockwell and J. Holmes, Electron. Dev. Lett. **23**, 300, (2002)

[9] A. Giussani, P. Rodenbach, P. Zaumseil, J. Dabrowski, R. Kurps, G. Weidner, and T. Schroeder, J. Appl. Phys. **105**, 033512 (2009)

[10] J.W. Seo, C. Dieker, A. Tapponier, C. Marchiori, M. Sousa, A.Ipsas, and A. Dimoulas, Microelec Eng. **84**, 2328 (2007)

[11] G. Saint-Girons, J.Cheng, P. Regreny, L. Largeau, G. Patriarche and G. Hollinger, Phys. Rev.B, **80**, 155038 (2009)

[12] G. Saint-Girons, G, P. Regreny L. Largeau L, G. Patriarche, G. Hollinger Appl. Phys. Lett. **91**, 241912 (2007).

[13] J. Cheng, P. Regreny, L. Largeau, G. Patriarche, P. Regreny, G. Saint-Girons, Appl. Phys. Lett. **94**, 231902, (2009)

[14] P. J. Dean, Phys. Rev. **157**, 655, (1967)

[15] G. Saint-Girons and I. Sagnes J. Appl. Phys. **91**, 10115, (2002)

[16] Z. H. Lin, T. Y. Wang, G. B. Stringfellow and P. C. Taylor, Appl. Phys. Lett. **52**, 1590, (1988)

[17] H. Lipsanen, M. Sopanen, J. Ahopelto, J. Sandmann and J. Feldmann, Jpn. J. Appl. Phys. Part 1 **38**, 1133, (1999)

[18] B.T.Seaman, L.D.Carr and M.J.Holland Phys.Rev.A 71, 033622(2005)

[19] G. Saint-Girons, C. Priester, P. Regreny and al. Appl. Phys. Lett. **92**, 241907 (2008)

[20] G. Saint-Girons1, J. Cheng, P. Regreny, L. Largeau, G. Patriarche, and G. Hollinger Phys.Rev.B 80, 155308 (2009)

[21] J. Cheng, P. Regreny, L. Largeau, G. Patriarche, O. Mauguin, K. Naji, G. Hollinger, G. Saint-Girons, J. Cryst. Growth 311, 1042, (2009).

AUTHOR INDEX

SUBJECT INDEX

Printed in the United States
By Bookmasters